Greener Pastures
On Your Side Of The Fence

ABOUT THE AUTHOR

Bill grew up on a dairy farm in Wisconsin, obtained a B.S. in Zoology from the University of Wisconsin, and served as a Peace Corps Volunteer in Chile. After his Peace Corps service, he returned to graduate school at Wisconsin with his new Chilean wife, Lita, and obtained an M.S. in Soil Science and a Ph.D. in Agronomy. He did part of his M.S. research and all of his Ph.D. research in Rio Grande do Sul, Brazil. While there he heard about Voisin grazing management, but never saw it being used successfully. After completing the degree requirements, they returned to Rio Grande do Sul to work on a United Nations project. After that, Bill and Lita moved with their daughters, Michelle and Nicole, to Oregon where they lived while Bill worked on an experiment station for Oregon State University.

Bill and Lita now farm 25 acres of land in the Green Mountain foothills of Vermont's Champlain Valley. It was here in 1981, while grazing dairy heifers and trying the Voisin pasture management method, that Bill realized the dramatic improvements in plant and animal production that result from using Voisin controlled grazing management. By 1986 the productivity of their pasture had quadrupled, with no other inputs besides observation, Voisin grazing management, and fencing. Bill also teaches courses and does research on pasture management, forage production, weed/crop ecology, and sustainable agriculture at the University of Vermont.

Greener Pastures On Your Side Of The Fence

Better Farming With
Voisin Grazing Management

by Bill Murphy

Second Edition

Arriba Publishing, Colchester, Vermont

Copyright ©1987, 1991 by Bill Murphy
All rights reserved.
    First Printing 1987
    Second Printing 1989
    Third Printing 1991, revised

Library of Congress Catalog Card Number: 86-90583

Published by: Arriba Publishing
                  213 Middle Road
                  Colchester, Vermont 05446
                  Phone: 802/878-2347

Printed in the United States of America

Library of Congress Cataloging-in-Publication Data.
Murphy, Bill
  Greener Pastures on Your Side of the Fence
  Better Farming with Voisin Grazing
  Management
  Includes index.
1. Pasture Management - Handbooks, manuals
2. Grazing Management - Handbooks, manuals
3. Livestock Production - Handbooks, manuals

ISBN 0-9617807-1-1 Softcover

To Lita

¡Nos hemos divertido harto!

¡Y seguimos pasándolo bien, mejor que nunca!

Acknowledgments

This book results from all of my experiences up till now. So everyone who has touched my life in one way or another has contributed to it. In particular, thank you Joe and Dolores Murphy for raising me on a farm; without that, I may not have developed a farmer's perception, sensibilities, and common sense. And thank you Caroline Alves, Doug Flack, Alice Pell, Dan Patenaude, Jeanne Murphy Patenaude, and Henry Swayze for helping to make this more useful by reviewing the manuscript, asking questions, and suggesting changes. And Murray Thompson, you asked some hard questions that forced me to get it right!

In the first edition I mentioned the pasture work that "we" had done in Vermont, but I didn't tell who "we" were. Besides myself, "we" included David Dugdale, who worked with and helped me for several years doing research at the University of Vermont. The other person was John Rice, who was the first Extension Agent in Vermont to see the value of feeding livestock on well managed pasture instead of unnecessarily in confinement. John risked his reputation in persuading three dairy farmers to control the grazing of their cows and participate in a study to measure pasture forage quality and yield. Without David and John this book wouldn't exist. David, I hope this isn't the Dugdale tweak!

Contents

Preface

In the dark forest a berry drops:
The sound of water.
                              Basho

The above words create a beautiful scene complete with sound in the mind's eye. On another level they describe exactly how I felt about this book when I first wrote and published it several years ago. This book was like that berry dropping into the large pond of agricultural writings and practices. In this book I describe a method that I know can help farmers to farm more naturally, efficiently, and profitably. The ripples from this book and others are spreading, but most farmers still are unaware of the method. Permanent pastures in humid-temperate regions of North America generally still are a neglected resource, producing far below their potential the way they usually are managed. Clear guidelines such as those presented here are needed for proper management of pastures in these regions.

Vague management suggestions, such as "Rotational grazing involves grazing each of a series of pastures in rotation and then moving animals to the next.", or "Rotation every 6 to 10 days of three or more pastures will

11

allow the plants to recover.", actually can do more harm than good. Following these kinds of general statements, farmers divide pastures into two or three parcels and pride themselves on it, only to have their animals run out of pasture forage in July or August. The pastures then get blamed for producing so little. But it isn't the pastures' fault! Low livestock productivity on pasture usually is due to extremely poor management of a forage crop in a pasture situation.

Understandably, many people in the United States question the value of rotational grazing. This attitude partly reflects conclusions about research results that showed only an 8- to 10-percent gain in livestock productivity from rotational grazing, compared to continuous grazing. These differences were too small to be statistically significant, and researchers concluded that rotational grazing isn't worthwhile.

But if looked at in another way, the insignificant statistics indicate that the rotational grazing designs that were tried didn't adequately meet the plants' needs, and therefore weren't really different enough from continuous grazing in their effects on pasture plants. The rotational grazing designs all had a basic defect that prevented their success: they didn't take into account the need to increase the recovery period between grazings as plant growth slows as the season progresses. They also didn't sufficiently limit the period that animals grazed a paddock at any one time.

The problem of pasture management in the United States has been dealt with in essentially two ways. One way has been through pasture renovation. This is an expensive process in which the sod is killed through use of herbicides or repeated cultivations, lime and fertilizer are applied, the seedbed is prepared or a sod-seeder is used, and a mixture of selected legumes and grasses are seeded. In short, an attempt is made to deal with permanent pastures as if they are field crops. But except for machine harvesting of excess forage, most field-crop technology can't be transferred to permanent pastures, even with major modifications. A permanent pasture is not the same kind of ecological environment as a field crop.

The final instruction in this renovation process has been: "Remember to change the grazing management or the pasture will become the same mess that you had before renovating." But no one ever stated exactly what that management change should be, so of course pastures didn't remain improved for very long.

Renovation of permanent pastures on much of the land in North America is economically or physically impossible anyway, because of steep slopes, shallow, rocky soils, ledge outcroppings, boulders, and brush. The land is considered to be marginal and is in permanent pasture because it is unsuitable for field crops. Besides, good evidence indicates that renovation isn't necessary. Grazing management of the pasture must be changed first, not last. Nothing else may be needed.

The other way that people have dealt with the problem of pasture messes has been to avoid pasturing or pasture management, through year-round feeding of stored forage to livestock in confinement. Some farmers continue to pasture animals in this system, but the pastures generally are considered to be mere holding or exercise areas that have little or no feeding value. Stored forage is fed year-round, in addition to feeding daily green-chopped forage during the growing season.

Although year-round confinement feeding did eliminate the need to manage pastures, it brought on many problems of its own, some much more serious than poorly managed pastures ever presented. Mastitis, hoof rot, other diseases, and breeding problems are much more prevalent in confined animals than among animals that graze during most of the year. All of these are symptoms

13

that something is wrong, and all increase production costs.

One of the biggest problems with confinement feeding is that it involves high capital investment in facilities and equipment and large amounts of purchased supplements. Feeding in confinement costs two to six times more than when animals graze their forage.

Unlike other businesses, farmers usually can't pass high production costs on to consumers. Farmers, especially in dairying, are now caught in a cost-price squeeze situation, in which higher operating costs together with lower prices for their products have resulted in greatly reduced profit margins.

Supposedly the large investment was made to save labor and increase profits, but it has turned out that farmers feeding year-round in confinement work a lot more, have lower profit margins, and consequently have a poorer quality of life than farmers who feed their livestock on well managed pastures during most of the year. Having grown up on a dairy farm, it's hard for me to understand why anyone in their right mind would choose to extend winter confinement feeding year round. We always counted the days until those cows could be turned out on pasture in the spring!

This unnecessary and expensive year-round confinement feeding is one of the reasons why many American farmers are experiencing financial difficulties that are driving many into bankruptcy and off the land. The work, debt load, and stress associated with year-round confinement feeding also make farming less attractive to farm children, and they look for other ways to make a living. Fewer farmers on the land results in serious social problems of deteriorating rural communities and landscape.

Farmers who remain come to prefer their neighbor's land to their neighbor. It's sad to see small family farms transformed into large farms with corn fields surrounding the previous owners' houses and stark silos. Have you noticed how small-farm barns tend to burn down after they're taken over by enlarging farms?

Many small farmers buy more new and used farm equipment, vehicles, and other supples than a few large farmers do. As numbers of farmers decrease, less is bought in town, and businesses earn less. Because of decreasing local employment opportunities, both farmers and townspeople move away to work in the cities. The town's services of churches, schools, hospitals, and libraries are needed less and less, consolidated, and finally closed.

Another serious problem to society as a whole is the environmental pollution resulting from high levels of pesticides and fertilizers applied to produce feed without crop rotations, as is generally done to support year-round confinement feeding of livestock. Also, more animals are kept than otherwise could be supported by the farm unit without purchased feed and supplements. The resulting excess manure applied to the limited land area of a farm overwhelms the soil environment, and leached nutrients and pesticides pollute ground water, streams, rivers, lakes, estuaries, and bays. Consequently, the Environmental Protection Agency has identified agriculture as the largest nonpoint source of surface water pollution! Little or no crop rotations, plus no animals pastured on the land, plus applying excessive amounts of manure, fertilizer, and pesticides, equals sick farmland. Symptoms of sick farmland include unusually severe pest problems and the need to apply large amounts of fertilizers and pesticides to maintain crop yields.

This lack of crop rotations and reliance on monoculture cropping of corn and other row crops is the most serious, though least obvious, defect of the year-round confinement practice. Quietly and gradually soil organic matter and humus have been decreasing under monoculture cropping, until the amount of soil erosion

15

occurring now equals or is worse than that of the dust-bowl years. Civilizations depend on farmers producing more food than the farmers need. Farming depends on the soil. Other civilizations have disappeared because they destroyed their soil base.

One solution to these problems may lie in better use of pastureland. It is an enormous resource that, together with rangeland, covers 47 percent (887 million acres) of the total area of the continental USA.

Development and efficient use of the pasture resource could benefit farmers in many ways. Feeding livestock on pasture can cost six times less than it does to feed them in confinement. Lower production costs with pasture feeding can accrue from: decreased grain concentrate, cropping, harvesting, storage, feeding, and manure handling costs; less machinery use, repair, and fuel consumption; lower labor requirements; and sale of surplus feed. Dairy cows fed on pasture have fewer herd health problems, thereby lowering veterinary costs, labor needs, and milk loss. Milk from pasture-fed cows has lower somatic cell count than milk from cows fed in confinement, which results in a premium price received for milk produced on pasture. These factors can increase farm profitability, while lowering labor requirements for the farmer, which together may improve quality of life for the farm family.

Society may benefit in various ways from widespread livestock feeding on pasture and improved farm family quality of life. Because more farms may remain in business due to higher profitability, and more farm children may go into farming because it is perceived as a desirable occupation again, the rural landscape will be maintained and rural communities will be rejuvenated. In addition, feeding livestock on pasture can result in less row-cropping, because livestock can obtain all or most of their feed from pasture during most of the year. In this respect, substituting pasture for corn, especially on sloping

16

soils, could significantly decrease soil erosion and environmental pollution from pesticides, fertilizers, and manure. Decreasing the amount of corn grown could alleviate some water pollution problems, because the 80 million acres of corn presently grown receive 55 percent of herbicides (220 million lb), 44 percent of insecticides (9 million lb), and 44 percent of nitrogen fertilizer (5 million tons) applied each year in the United States. Depending on slope, rates of soil erosion from corn fields far exceed soil formation rates.

All in all, an end result of the shift to year-round confinement feeding is that pastures have been practically eliminated from American farming experience for a generation, and pasture management became a lost art in the United States. It is just as well, though, because pastures never were properly managed in this country, and the old ways are better forgotten. Also, new electric fence chargers and fencing materials developed mainly in New Zealand are available now to make a return to pasturing in the United States relatively easy.

New Zealand's highly productive and profitable agriculture depends almost entirely on permanent pastures, using grazing management initially defined by Andre Voisin of Normandy, France. Voisin saw pastures differently than other people did. He felt that if we take care of the pasture plants and soil life, they will take care of the grazing animals. And that is just what happens. For example, New Zealand farmers feed about the same number of cattle as there are dairy cows in the United States, and seven times more sheep (plus 1 million each of deer and goats), but they do it on a pasture area the size of Wisconsin! And without grain supplements!

In the United States some people refer to the grazing management method used in New Zealand as intensive rotational grazing, but this may be misleading, and there's a lot more to it than what that may imply. Voisin called

17

his method "rational grazing". Rational grazing can mean two things: the thinking way of grazing management, or a method of rationing out the forage. When I first heard of the method in Brazil, it was called Voisin grazing management, and I proceeded to call it that. Since Voisin defined rational grazing, I still think it is appropriate to use his name, but it may have added to the confusion of terms that have come into use.

Other terms being used to describe this grazing method include short duration, controlled, or planned grazing management. Short duration is somewhat confusing, since it emphasizes grazing period in a paddock, when actually recovery time between grazings is more important. Voisin mentioned controlled grazing in his book, but preferred to use a new term to describe his definition of the basic elements of the method in a way that he felt would be useful to farmers. Planned grazing management was coined by Allan Savory to describe the method mainly when applied to rangeland. It doesn't make any difference what you call the method, just so what you do in practice includes the key elements of adequate plant recovery periods between grazings, and short grazing periods with high stocking density. Whether you're grazing pasture or rangeland, planning certainly must be part of your management. Here I'll use all or parts of the term "Voisin controlled grazing management", and leave the word "planned" for rangeland grazing.

(If you're grazing livestock on rangeland in a brittle environment, you need to use planned grazing management. It involves planning for specific recovery periods between grazings, monitoring regrowth of severely grazed plants, and using herd effect to break soil crust to allow water penetration and seedling establishment. All environments, regardless of total rainfall, fall somewhere on a continuous scale from brittle to nonbrittle. A completely brittle environment is

18

characterized by: 1) unreliable precipitation regardless of volume; 2) poor distribution of atmospheric moisture through the year; 3) mainly chemical (oxidizing) and physical (weathering) "decay" of old plant and animal material that's generally slow and from upper parts downwards; 4) very slow successional development from bare and smooth soil surfaces, often stopping at algal capping, and on steep slopes not even reaching algal stages; and 5) with a lack of adequate physical disturbance for some years, successional plant communities become simpler, less diversified, and less stable. For more information contact: Center for Holistic Resource Management, PO Box 7128, Albuquerque, NM 87194; telephone: 505/242-9272.)

Voisin controlled grazing management can be used in North America just as well as it is in New Zealand and elsewhere, to allow pastures to reach their full productive potential. The principles are the same, but adjustments are needed to fit local conditions.

What I present here may seem to be slanted toward permanent pastures, rather than pastures that are part of a crop rotation (rotational pastures), but this is only because most farmers right now want to increase the productivity of their so-called marginal land, which usually is in permanent pasture. Also, this "marginal" land is ideally suited for ruminant livestock production. Ruminants can use forage grown on such land to produce meat, milk, and wool, and not compete with production of other human food and fiber on arable land. Everything in this book applies to rotational pastures, but differences certainly exist. For example, earthworm numbers in the soil of a pasture rotated with field crops will be much less than those in an adjacent permanent pasture, even though both are managed according to the Voisin method. I point out important differences if they relate to management, so that they can be taken into account when dealing with

19

rotational pastures.

As more and more farmers return to proper 5- to 7-year crop rotations (as they surely must do before monoculturing or two-crop sequences destroy the soil and ruin the environment), pastures will become essential parts of crop rotations again. I hesitate to call these kinds of pastures rotational pastures, because it results in such terrific confusion of terms, and implies that these pastures are being rotationally grazed, when actually they might be continuously grazed. So why don't we now, before pastures commonly become part of crop rotations again, agree to use the sensible British term, "leys", for such pastures? In this book, at least, leys are pastures that are part of arable crop rotations.

It's unlikely that prices for agricultural products will increase, at least for awhile, but you can improve your profit margin now, by reducing production costs. This is a book for farmers who want to increase the profitability of their farms and decrease their work load, while properly and responsibly caring for the land, and the plants and animals that live on it. It isn't evidence in support of a method that I want you to believe in. Believing is not knowing! This book explains a technique enough so that you can try it and experience it for yourself. Then you will know that it works!

Probably the greatest benefit arising from using pastures as valuable parts of farms, isn't in terms of dollar profits, but the peace of mind and mellowness that naturally develop in farmers as they get back in touch with the land, and stop running to feed confined livestock. Pastures are fun! So shut off that 160-horsepower articulated tractor, climb down from that air-conditioned, stereo-rocked cab, and take a leisurely walk around your pasture. Lie down under an old tree in the pasture. Chew on some clover flowers. Enjoy this wonderful world we live in!

1

Pasture Plants

If I look closely I can see
The shepherdspurse
Blooming beneath the hedge.
                    Anonymous

Permanent pastures are amazingly complex environments, compared to row crops such as corn. Twenty to
thirty plant species can be present, especially when the
pastures have not been grazed intensively. In a row-crop
situation one species, the crop, is present plus any weeds
that managed to escape being killed by tillage or herbicide
application. Even if the row crop is very weedy, still only
five to ten weed species usually are present. Leys are
actually more similar ecologically to row crops than to
permanent pastures, in that leys have at most two or three
species each of legumes and grasses, plus a few "weed"
species, and are periodically plowed under, only to be
reseeded later. Overstocked, continuously grazed pastures,
whether permanent or ley, may only contain one legume
and one grass plus low-growing "weeds". These will be
plants that have or can assume a growth form close to the

21

soil surface below the animals' grazing height.

If you slowly walk through your pasture looking carefully, you will find a world that you may not have known existed. Take along field guides to grasses, legumes, weeds, and wild flowers so that you can identify the plants. And take a small shovel to see what is happening underground. Usually in pastures we see the forest and not the trees. Let's look at the trees!

As many as ten or more grass species may be present in an undergrazed permanent pasture. Among others, in the northeastern and northcentral United States these usually include quackgrass, Kentucky bluegrass, Canada bluegrass, timothy, bromegrass, orchardgrass, annual and perennial ryegrasses, tall fescue, and reed canarygrass. In extremely overgrazed pastures the only one present may be Kentucky bluegrass. All of these have different growth forms, growth habits, and carbohydrate reserve cycles.

Several legumes may be present, including alfalfa, red clover, white clover, vetches, and sweetclover. In overstocked, continuously grazed pastures, white clover probably will be the only legume present.

In low-lying areas of pastures, reed canarygrass, sedges, and/or rushes grow, depending on how wet the areas are. Sedges and rushes have feed value, but aren't very palatable to animals. A beneficial thing to do with areas growing sedges or rushes is to fence them off from the rest of the pasture, so that livestock can't walk through them, and they'll become refuges for birds and other wildlife.

A large number of broadleaf plants, such as plantain, dandelion, and chicory are present in pastures. The number and diversity depend on grazing management. More species in greater numbers will be present in undergrazed pastures. Fewer species and only low-growing ones, except for those that protect themselves from grazing (e.g. thistles), can survive in overstocked, continuously grazed pastures. Although this group of plants generally is looked down upon by people and called "weeds", grazing animals have a different perspective. Many "weeds", such as plantain and dandelion, are actually preferred by grazing animals over the grasses and legumes that humans have selected. Almost all "weeds", except plants that protect themselves from being grazed with spines, thorns, odors, or poison, are edible and nutritious for livestock in early growth stages. For example, dandelion forage that I analyzed for feeding value contained 15% dry matter, 22% crude protein, 72% total digestible nutrients, and 0.76 Mcal NEL/lb DM!

Another group of plants that you'll find include some of the oldest plants in existence: ferns, horsetails, and lichens. They are found especially along edges of woods in moist or shaded areas. These plants generally are good feed for livestock, except for some of the ferns, which are poisonous.

When you start controlling grazing be careful, especially in the beginning, so that you don't force your livestock to eat poisonous plants that they normally would not eat. It isn't that they can't eat any poisonous plants; a few might not harm them. The problem develops when animals eat large amounts of poisonous plants in proportion to nonpoisonous ones. To be safe, and especially if there are a lot of poisonous plants present, either remove the poisonous plants or exclude the areas containing them from the pasture.

Finally, the most obvious plants you will see, sometimes more than you want to, are bushes and shrubs. These can cover large areas of a pasture, especially when it has been understocked and continuously grazed. Bushes and shrubs seen from the side at our sight level may seem to be more dense than they actually are. Scattered bushes and shrubs actually block very little sunlight, help hold the soil in place, recycle nutrients that have been leached below pasture plant rooting depth, and provide browse for livestock. Some fix nitrogen. Check to see if the bushes and shrubs are far enough apart to allow sunlight through for grass and legume growth. If they're scattered enough, don't worry any more about them because they won't increase under Voisin grazing management. If they're growing too densely, thin them out until they cover about 30 percent of the area.

Nature always moves toward the most well adapted and stable group of plants in any situation (climax vegetation). For example, bushes and shrubs are an intermediate stage in the movement toward a tree climax in northeastern and northcentral United States. This region used to be covered by forests, and growing condition forces move it rapidly back to that stage, unless work is done to maintain it as grassland. Controlling the grazing of livestock is a way of doing that work while reaping the benefits from highly productive pastureland.

Understocked, continuous uncontrolled grazing actually helps to move pastures along to a climax vegetation. Animals select the most palatable plants, usually leaving tall-growing broadleafs, bushes and shrubs, and young trees. These then have less competition and can grow more quickly, and the climax stage is reached sooner, unless someone intervenes with saws and land-clearing equipment to start the cycle all over again. By controlling livestock grazing, you avoid this problem and expense.

If this walk through your understocked pasture occurs in mid to late summer, you will be greeted by a profusion of colorful flowers. People with highly developed sensibilities have said that colors of plants and flowers reflect cosmic forces. Accordingly, blue flowers reflect the influence of Saturn, Jupiter brings forth white or yellow flowers, Mars can be recognized in red flowers, and in green leaves we see essentially the Sun. In a pasture, certainly there is more than meets the eye!

In a way, it is a shame to manage a pasture more intensively, because the plants almost never flower under such management. But since plants reflect cosmic forces, a well managed pasture with its almost throbbing green color, only indicates that more Sun is being expressed than other influences.

Along with the decrease in flowering that occurs, the pasture plant community simplifies, as species that can't withstand intensive grazing disappear. Let's look more closely at plants likely to persist under Voisin controlled grazing management.

GRASSES

You will find grasses with two main growth forms in your pastures: bunch-type and sod-forming. Growth forms are one of the obvious things that affect how the grasses relate to other plants in the pasture and respond to grazing management.

Bunch-type grasses grow in bunches in the spot where the original seed germinated and established. Open space exists between the bunches and is filled with other plants.

Sod-forming grasses spread away from and around the spot where the original seed established. Many sod-forming grasses have underground stems (rhizomes) that produce roots and new shoots at various distances away from the mother plant. Other sod-forming plants have similar stems lying on the soil surface (stolons) that produce shoots and roots at nodes contacting favorable soil conditions. The new plants in both kinds of sod-formers eventually become independent of the mother plants, and produce new plants of their own. The rhizomes and stolons, along with the extremely finely divided, dense, moderately deep root systems common to most grasses, make these grasses especially valuable for soil and water conservation.

25

Growth habit is another obvious characteristic of plants that affects how they respond to grazing management. Low-growing grasses are well adapted to close grazing. They produce many side shoots (tillers) from their bases early in the season, but don't grow upright until ready to flower. This keeps their growing points below the level of grazing, and a thick, vigorous stand can be maintained under close grazing.

In contrast, tall-growing grasses are easily damaged by close grazing because their growing points that produce regrowth rise high above the soil surface, and are removed by close grazing. Depending on the grass and how much carbohydrate reserve it has, little or no regrowth may occur after close grazing early in the season.

Orchardgrass

Orchardgrass is a bunch-type grass with a moderately low growth habit, so it can withstand fairly close grazing. In fact it needs to be grazed intensively to keep shoots palatable to livestock; otherwise it tends to become coarse and very bunched, and is not readily eaten.

Orchardgrass is a valuable pasture plant that is well adapted to moderate soil fertility and low soil moisture conditions. It begins growth earlier in the season than associated legumes and recovers quickly after grazing. So pasture areas containing orchardgrass can and should be grazed early and frequently to keep it in a young leafy stage of growth.

Timothy

Timothy is another bunch-type grass that you're likely to find in your pasture. It's also a moderately low-growing plant, and stores carbohydrates in a swollen lower area (corm) of the shoot for use in producing regrowth. You can see or feel the corms by looking at or touching the plants just above the soil surface. Timothy probably is the best all around grass for palatability, persistence, establishment ease, yield, disease resistance, and tolerance of poor drainage. It tolerates everything but drought.

Kentucky Bluegrass

Kentucky bluegrass almost always exists in pastures because it can withstand even the worst grazing management, since its growing points remain very low to the ground. This same characteristic allows it to survive weekly mowings down to 1 inch in lawns. It is a strong sod-former that spreads by rhizomes. Kentucky bluegrass produces most of its growth during the cool parts of the season, and grows slower in midsummer. When managed well it produces high yields of excellent quality forage, and grows more uniformly throughout the season.

Bromegrass

Bromegrass forms a dense sod as it spreads throughout pastures by short rhizomes. It is an extremely winter hardy grass that tolerates hot, dry conditions. Bromegrass provides excellent early spring grazing, when its growth is mostly leaves. But because it grows upright and raises its growing points above the level of grazing in late May, the regrowth after the first grazing should be saved for a hay crop if possible, to help bromegrass persist well. Grazing bromegrass should stop when new shoots at its base (crown) begin to grow. Once the hay crop is harvested and the plants rested for about 25 days, grazing can begin again.

Reed Canarygrass

You will find reed canarygrass growing in the fertile, low,

moist areas of your pasture. It also is a sod-former that spreads slowly by short, thick rhizomes and forms heavy tough sod that is often bunchy. It produces most of its growth during the cool parts of the season, beginning very early in the spring. It has to be grazed early and kept well grazed down to less than 12 inches, otherwise it becomes unpalatable to livestock. Common reed canarygrass usually has poor palatability because it contains toxic alkaloids, but if livestock don't eat too much of it, they aren't harmed. New cultivars have been selected that contain less alkaloids. It is better feed than the sedges or rushes that would grow in the low areas if reed canarygrass wasn't there.

Perennial Ryegrass

If you live in growing zone 4 or colder, you probably won't find perennial ryegrass in your pasture unless you have seeded some of the cold-tolerant cultivars that recently have become available. This is an ideal pasture grass where it is adapted. Under high soil fertility (especially nitrogen) conditions, it produces large amounts of excellent quality forage. Perennial ryegrass is very persistent where it's adapted, and combines well with white clover, while withstanding intensive close grazing.

Tall Fescue

Tall Fescue essentially is a bunchgrass, although it has short underground stems. If it is kept closely grazed or mown, it produces a dense sward and tight sod. Tall fescue is used widely for soil conservation purposes, turf, and forage. It tolerates continuous close grazing, and is particularly valuable for spring, fall, and winter grazing. If cattle are removed in midsummer and forage is allowed to accumulate until the first frosts occur, tall fescue pastures can carry cattle through the winter. Autumn frosts improve the its palatability.

Quackgrass

Quackgrass is another grass that you will almost always have in your pasture. This isn't such a bad grass to have around; in fact quackgrass probably holds more North American soil in place than any other plant! Quackgrass has good feed value, but most of its growth occurs as rhizomes. Unless you're grazing pigs, that doesn't help you much. Quackgrass, however, competes strongly with other plants, especially legumes, and for that reason the pasture is better off without it. When grazing is controlled, quackgrass eventually disappears from the sward because it can't withstand such intensive grazing. Competition from vigorously growing grasses and white clover that are well adapted to controlled grazing also helps to rid a pasture of quackgrass.

Canada Bluegrass

Canada bluegrass is considered a weed mainly because it begins growth and flowers early in the season. The flowering stems are very unpalatable, especially to cattle, possibly because they are unable to bite through the stems. Since ruminants only have front teeth on the bottom jaw, they eat by grasping the plants with their tongues and mouths and tearing the plants off. When they do this with the flowering stems of Canada bluegrass and other plants, the stems probably cut their gums. So cattle avoid eating plants such as Canada bluegrass and any neighboring plants so they don't get bluegrass stems between their teeth. By beginning grazing early in the spring, flowering stems can be grazed off before they elongate very much and become too fibrous to be torn off. If it flowers before it can be grazed, mowing paddocks on the last day that animals are in them, allows the animals to eat the stems and other plants growing among Canada bluegrass.

LEGUMES

You should find at least scattered legumes in your pasture, no matter how poorly it has been managed, even though animals selectively graze legumes. If you don't find any legumes, a soil fertility or pH problem most likely exists. Legumes are extremely important to your pasture's productivity and feeding value of the forage that it produces. Although air contains 80 percent nitrogen, it is unavailable to most organisms. Legumes don't require soil nitrogen because they can use nitrogen from the air. Rhizobia bacteria that associate symbiotically with legumes in nodules on the plant roots, combine (fix) atmospheric nitrogen so that it can be used by the legumes and, consequently, by all other plants and animals. This nitrogen ultimately becomes available to associated grasses through urine and manure excreted by grazing livestock, and decomposition of legume nodules, roots, and shoots.

As though fixing nitrogen was not enough, legumes also improve forage quality of plant mixtures. Compared to grasses, legumes contain less fiber, have a higher ratio of soluble to insoluble carbohydrates, have double the mineral content, and are higher in protein.

White Clover

White clover usually will be the main legume sustaining your pasture's productivity. It withstands the most intensive grazing, fixing large amounts of nitrogen while producing high yields of nutritious forage. Most likely, after your pastures have been managed well for a few years, white clover will be the legume present in largest amounts.

White clover grows rapidly, spreading by creeping stolons lying on the soil surface. Kneel down and follow a white clover stolon with your fingers. Notice how every few inches roots have formed at nodes. These rooted spots eventually become independent plants that will then spread on their own. Carefully dig up one of the larger rooted plants and separate the soil from the roots. On the very finest root hairs you should be able to see pinkish white round formations. These are the nodules in which nitrogen from the air is being fixed by rhizobia, making it available to all other living organisms, including yourself.

While you are down there, carefully dig and loosen more clover roots and some grass roots from the soil. Notice how intimately the roots associate with the soil. With an electron microscope you would be able to see that life is continuous between plants and surrounding soil; no distinct line exists.

Since it grows so low to the ground, white clover withstands grazing very well. No matter how heavily grazed, some leaves always remain below the level of grazing. This enables white clover to continue photosynthesizing and obtain the energy needed for regrowth. The white clover forage that animals eat is all leaves, petioles, and flowers. So the forage has extremely high feeding value all season.

White clover grows on a wide range of soil conditions, including acid soils down to about pH 4.5. White clover does best on fertile, moist soils, and doesn't do well on light sandy soil because of its shallow root system. It grows well at a soil pH of 5.5 to 6.0, because the rhizobia that infect clovers and fix nitrogen function well in that pH range. So it isn't necessary to lime white clover pasture soils to pH levels higher than 6.0. Healthy looking white clover plants with pink nodules indicate that soil conditions are favorable or can be improved without much difficulty or expense.

Red Clover

Red clover is more deeply rooted than white clover and, consequently, can grow on a wider range of soil moisture contents. Otherwise it grows well under soil conditions similar to those favored by white clover. Because red clover grows more upright than white clover, it can't survive intensive grazing as well as white clover can. Red clover probably will persist, however, especially in areas that you set aside in the spring for machine harvesting.

Alfalfa

Alfalfa is the most deeply rooted legume that we use and can grow on the driest soils. In fact, it doesn't do well at all on soils that are too moist or poorly drained. Alfalfa is a very productive legume under conditions where it is well adapted and well managed. Because the rhizobia that infects alfalfa requires a soil pH of 6.0 to 7, you probably won't find it in your pasture unless its soil pH is in that range. Most soils of permanent pastures in the Northeast have pH levels below 6.0.

Besides its requirements of higher soil pH and good soil drainage, alfalfa can't withstand as intensive grazing as the clovers. Because of its growth habit and carbohydrate reserve cycle, alfalfa regrows mainly from stored carbohydrates, not from leaves left after grazing. So alfalfa can be grazed closely, but not frequently. Alfalfa may not persist at all if grazed in the first and second spring rotations. At that time alfalfa's carbohydrate reserves are low, and it needs several weeks to replenish them through photosynthesis during the first growth of spring. Alfalfa seems to persist well under controlled grazing if the first alfalfa crop is machine-harvested in June. After resting the alfalfa for 25 to 30 days, it can then be included in the grazing rotation without severely damaging the plants. This is because at that time the rest or recovery periods between grazings are long enough to meet alfalfa's needs.

Birdsfoot Trefoil

You might find birdsfoot trefoil in your pasture; it's hard to know where this legume will establish and grow well. Birdsfoot trefoil can grow in extremely adverse soil conditions. If your pasture soil is a clay that floods in spring and bakes as hard as concrete in summer, trefoil may do well there. It also grows well along roads where salt is spread during winter. While it will grow on poorly drained, acid, droughty, infertile soils, trefoil produces best on heavy fertile soils at pH 6.5. Besides fixing nitrogen and producing large amounts of nutritious forage, birdsfoot trefoil has the added advantage of not causing bloat in grazing animals.

All of this indicates that trefoil should be a great legume to have in your pasture. Trefoil maintains a low level of carbohydrate reserves during the growing season, regrowing mainly from solar energy captured by photosynthesis in existing leaves. This means that enough postgrazing trefoil residue should be left so that leaves remain to continue photosynthesizing for regrowth. Using prostrate cultivars such as Empire help to leave a long residue with leaves. By keeping the grass-trefoil sward from growing more than 6 to 8 inches tall before grazing, you can prevent lower leaves on trefoil from being lost due to shading, thereby helping it recover quickly after grazing.

Other Legumes

Alsike clover and vetches are two other legumes that you

may find on this leisurely walk through your pasture. Alsike resembles red clover in growth habit, and produces better than either red or white clovers on wet soils under cool conditions, and in lower soil pH levels. The vetches are viny legumes that climb up neighboring plants and persist in understocked pastures.

REFERENCES

Heath, M.E., D.S. Metcalfe, and R.F. Barnes (eds.) 1973. *Forages.* Iowa State University Press, Ames. 755 p.

Langer, R.H.M. 1973. Grass species and strains. *In*: R.H.M Langer (ed) *Pastures and Pasture Plants.* A.H. and A.W. Reed, Wellington, New Zealand.

Savory, A. 1988. *Holistic Resource Management* . Island Press, Washington, DC. 565 p.

Semple, A.T. 1970. *Grassland Improvement.* Leonard Hill Books, London. 400 p.

Smetham, M.L. 1973. Pasture legume species and strains. *In*: R.H.M. Langer (ed) *Pastures and Pasture Plants.* A.H. and A.W. Reed, Wellington, New Zealand.

Smith, B., P. Leung, and G. Love. 1986. *Intensive Grazing Management.* The Graziers Hui, Kamuela, Hawaii. 350 p.

Smith, D. 1981. *Forage Management in the North.* Kendall/Hunt Pub. Co., Dubuque, Iowa. 258 p.

Steiner, R. 1977. *Agriculture.* R. Steiner Press, London.

Pasture Ecology

In the landscape of spring
There is neither high nor low;
The flowering branches grow naturally,
Some long, some short.

Zenrin

Changes in pasture plant populations occur because some individual plants are better adapted to their surrounding conditions and leave more descendents than others. This determines individual fitness. Aspects of individual fitness influence how plants grow, reproduce, and spread.

Within a pasture plant community or sward, individual plants gain an advantage over their neighboring plants in various ways. The kinds of activities that give an advantage usually are selfish qualities that include taking up more nutrients than needed, transpiring more water than needed, and producing a denser canopy than needed to intercept an optimum amount of light. So a plant actually gains in fitness if its activity harms its neighbors, since fitness depends only on leaving more descendents.

Unfortunately, many of the things that favor individual plants of one species over another, or plants of one genetic makeup over another in leaving more descendents, work against the cooperative community aspects that matter to farmers. (Many smaller plant communities exist in every pasture sward, depending on several things including soil type, slope, aspect, and water-holding capacity.) The selfish qualities important to the individual plant may produce a pasture sward that's quite different from what's needed for good forage production. This is because the fate of an individual plant depends on its immediate neighbors, which because of their demands on environmental resources, directly influence its wellbeing. The processes involved concern fitness at the level of individual plants, not forage productivity measured at the community level.

To understand the nature of the forces that influence individual plants within a pasture sward, we need to consider what happens in the sward from the plant's point of view. If we know how individual plants experience their local environments, we might be able to influence that experience to favor improved forage production.

COMPETITION

As long as all necessary growth factors are adequate to meet the needs of all plants within a sward, competition doesn't occur, and no individual plants are favored over others. But as soon as the immediate supply of a single factor becomes less than the combined needs of the plants for that factor, competition begins. This is the point when individual plants begin to gain or lose in fitness, and changes begin to occur in the sward.

Competition among plants occurs for water, nutrients, light, carbon dioxide, oxygen, and means of pollination and seed dispersal. Other things such as temperature and humidity certainly affect plant growth and reproduction, but are not limited in supply, and therefore not strictly involved in competition. Except for some root crops (e.g. carrots), competition for space rarely occurs. Usually there is plenty of space for more plants, leaves, branches, and roots.

Most of the things that plants compete for exist as a pool, which the individual plants draw from. If the pool of a certain growth factor is limited, the successful competitor is the plant that draws on the limited factor most rapidly from the pool, continues to draw on it when the supply is low, or is able to draw on it when other plants can't because the pool is too low. A successfully competing plant usually is one that quickly uses its immediate supplies of growth factors, and then extends into neighboring areas by growing more roots and leaves.

Almost all of the advantages gained by successfully competing plant species can be described by two words: amount and rate. These advantages include greater carbohydrate storage in seed or roots, more rapid and complete germination, earlier growth start in spring, faster growth of tops and roots, taller and more branching stems, deeper and more spreading roots, more tillers, more flowers, and larger leaves.

Any or all of these properties help individuals survive and leave more descendents. All of these traits express genetic and physiological backgrounds that enable the plants to take up water or oxygen from wet or dry soils, place leaves advantageously for light interception, take up nutrients that are less available, and adapt in growth and development to changes in climatic conditions.

GRAZING LIVESTOCK

In addition to all of the above influences, grazing

37

livestock greatly affect the composition of pasture plant communities. Animals always cause a pasture to be a more complex mixture of plants than it otherwise would be. This is because animals graze selectively and in patches, and the effects vary in time and in space. Animals also drop their manure and urine in patches, which affects some plants more than others. Besides walking, running, and jumping on a pasture, animals sit, lie, scratch, and paw on it. All of these things result in a sward containing a wide variety of plants adapted enough to survive their different local conditions.

COMPETITION FOR LIGHT

If it were possible to separate all the components of your pasture, sunlight certainly would be ranked as the most important and having the greatest influence on its botanical composition and yield. (Of course, nothing can really be separated from anything else: everything interrelates or "goeswith" everything else.) Now you would think that there would be plenty of light for all the plants in a pasture, so why is there competition for it?

The reason is that light decreases as it passes through a foliage canopy, because the leaves intercept and absorb it. If light is reduced too much, growth and development of shaded plants become slower. The effect of shading depends on the plant's stage of development and how much it is shaded. Another characteristic of shading is that the reduced light level in the canopy isn't the same over all plants in the canopy, or even all parts of single plants. Bottom leaves are shaded more than top leaves, and low-growing plants have more of their leaf surface shaded than tall-growing plants. Pasture plant canopies are surprisingly dense: a couple of inches of pasture canopy intercepts the same amount of light as several feet of forest canopy!

Competition for light occurs whenever a part of a plant interferes with the light supply falling on another plant or part of the same plant, and the photosynthetic rate of the shaded leaf area decreases. The only times when there is little or no competition for light are when plant growth first begins in spring and in the early stages of regrowth after grazing.

Two basic aspects of competition for light greatly affect the performance of pasture plants, and are the main reasons why Voisin controlled grazing works: (1) Light is not like a soil nutrient that remains in the soil until used. Light must be used instantaneously as it passes through the canopy, or it is lost forever. (2) The position of leaves within a pasture plant canopy is extremely important. Leaves that overtop other leaves gain a competitive advantage for light.

Several properties of light and plant canopy combine to influence competition for light, and they change all the time. These include the sun's angle of elevation, whether the light radiation is direct or diffuse, density of the plant canopy, leaf angles, and light reflection, absorption, and transmission characteristics of leaves. So plants that persist in a pasture sward must contend with a daily and seasonally changing light environment, especially during the regrowth period after grazing, plus withstand the effects of losing most of their leaf surface during grazing. It is amazing that any plants survive such treatment! Is it any wonder that pasture plants need increasing recovery periods during the season?

Effects of Management

Plant height is extremely important in competition for light. Even small height differences give considerable competitive advantages or disadvantages, especially during leaf development or elongation. Anything that

39

influences a plant's height or ability to shade in relation to its neighbors, determines its relative competitiveness. Pasture managers can influence two things that directly affect the relative competitive abilities of pasture plants for light: grazing frequency and soil fertility.

Clover and grass contents of swards mainly reflect the variation in their abilities to compete for light. For example, increased nitrogen from urine or fertilizer stimulates grass growth, which increases its ability to shade clover. More frequent grazing prevents shading of the clover, and its competitive ability increases relative to the grass.

Weedy pastures result from poor management that gives weeds a competitive advantage for light. For example, weedy plants with long wide leaves suppress shorter plants such as white clover by shading, if a pasture is understocked and grazed infrequently. But this growth habit gives no competitive advantage under frequent grazing with high stocking density. Allowing selective grazing of shorter desirable plants, gives uneaten tall-growing broadleaf weeds a great competitive advantage for light that enables them to dominate the sward. Infrequent or continuous understocked grazing are the main reasons for the weedy messes typical of most North American pastures, mainly because they give weeds the competitive advantage for light over clovers and grasses

WATER

Pastures usually depend on rainfall for water. When rainfall is irregular and soil water becomes limited, plant productivity and persistence can be severely affected.

One of the first consequences of dry conditions is that plant leaf area decreases. Because less photosynthesis can then occur, this quickly results in a reduced growth rate. Plants that can continue taking up water from a drying soil immediately gain a competitive advantage over plants that are unable to do so.

As conditions become drier, plants increase root growth to take up water from deeper in the soil, and restrict leaf growth even more to decrease water loss through evapotranspiration. Those that reduce their leaf area first or most, gain by decreasing water loss, but lose in competitive ability for light. Plants that can increase their root system, without decreasing their leaf area, are the ones likely to dominate a pasture sward during and right after a dry spell.

The productivity and persistence of pasture plants under drought stress also depends on the levels and availability of soil nutrients before and during a dry period. When the water supply is limited, less nutrients can be taken up by plants. Much less nitrogen is fixed by legumes, probably because their rate of photosynthesis decreases, and nitrogen-fixing bacteria in the legume root nodules become starved for carbohydrates.

One reason why plant nutrition and water supply are so closely linked is that nutrient levels usually are highest near the soil surface, which is the first layer to dry out. Although plants may have roots in the deeper and wetter parts of soil, the low amount of nutrients in subsoil and unavailability of nutrients in dry surface soil, may limit plant growth and forage yield more than the lack of water itself.

So dry conditions adversely affect plants in two main ways: 1) the shortage of water itself restricts leaf area, and 2) nutrient deficiencies develop that further limit leaf area. Both of these directly affect plants' ability to compete for light. For these reasons, pastures need especially good grazing management during and after dry spells to maintain a desirable plant composition in the sward. Always increase recovery periods at these times to allow the plants to recover and survive.

TEMPERATURE

As important as water is in a pasture plant community, temperature is even more important. The natural distribution of plants is determined by their adaptation to major climatic and soil aspects of the environment. Temperature limits the distribution and diversity of plants that can grow and develop in a region. As temperature goes above or below 68 degrees Fahrenheit (20 C), it becomes more important in plant productivity and persistence.

Since temperatures fluctuate all the time, regularly or irregularly, plants must be buffered or be flexible enough to adjust to short-term temperature changes. They also must have adaptive systems that synchronize growth and reproduction with daily and seasonal temperature cycles. Plants generally accommodate temperature extremes by becoming dormant in midsummer or winter, and develop temporary cellular resistance (hardiness) to heat or cold. Many use temperature changes to begin germination and reproduction.

In the Mediterranean region, cool winter and early spring are the most favorable growing seasons; midsummer drought limits plant growth. Plants in this region are adapted to growing actively at moderately low temperatures, but tend to become dormant at high temperatures, especially if water becomes limiting. Since many of the grasses that we use in pastures come from this region, our pasture productivity follows their growth curves. A lot of forage is produced during cool moist weather of spring and fall, but less during hot dry weather of midsummer.

Grazing management must take into account this variation in plant growth rate during the season. The main requirement of the Voisin method is to vary rest periods according to plant growth rate, to allow plants enough time to regrow and recover from the effects of grazing.

PLANT ROOTS

Probably because they are out of sight, hardly anyone considers the root systems of plants when managing pastures. But, just as plant tops are affected by selective grazing of variable intensity, roots are also influenced. Since water and nutrients are taken up by roots and nitrogen is fixed in the roots, any beneficial or adverse effects on the roots are important to pasture productivity.

Root temperature is always close to soil temperature. If root temperature rises above or falls below the optimum for a plant, growth of that plant slows or stops. The plant immediately is at a disadvantage in relation to its neighbors, if they have different genetic traits that enable them to withstand or use higher or lower root temperatures better.

Soil moisture not only affects forage yield, but also how carbohydrates produced in photosynthesis are used. For example, as soil dries, carbohydrates move to the roots to support increased root growth, as plants search for more water.

Frequent grazing to low plant residual or pasture mass without adequate recovery periods, results in reduced pasture productivity, partly because the amount of roots that plants can maintain decreases. After every grazing, regrowth is supported by carbohydrates that had been stored in stem bases and roots. So the amount of carbohydrates available from storage, influences how much and how fast plants can regrow after grazing.

The rhizomes and stolons of some pasture plants give them a competitive advantage, especially in resisting treading by grazing livestock. Plants with deep taproots, (e.g. alfalfa and some weeds) can compete successfully

43

against shallow-rooted plants (e.g. white clover and grasses), especially in drier soil conditions.

Weeds frequently flourish when growth of desirable plants is depressed by adverse conditions or poor grazing management. When desirable plants are grazed in preference to weeds, root systems of the desirable plants become restricted, and the weeds gain in competitive advantage.

It is interesting to note that about half of pasture plant growth is unavailable to livestock, because it is in the roots. The proportion of plant tops to roots increases if soil nutrients are present in adequate amounts, because the plants don't need to develop such extensive root systems to find and take up the nutrients.

SOIL ORGANISMS

Botanical composition and productivity of a pasture sward reflect its soil. Quality of a soil depends on the life within it. So in trying to improve pasture productivity, we should concentrate on ways that favor development of soil life.

Microorganisms

Pasture soils contain some of the highest root concentrations of all crops. And there are 20 to 50 times more bacteria and fungi in soil near plant roots (rhizosphere) than in soil away from roots. So the microbiology and chemistry of pasture soils actually concern the rhizosphere, rather than soil alone. In the rhizosphere chemical soil properties and microorganisms are affected by living plant roots, and vice versa.

Chemistry of the rhizosphere differs from soil away from plant roots because of movement of ions to roots, uptake of nutrient ions by roots, and release from roots of balancing ions and soluble organic materials. These materials include organic acids, sugars, and amino acids. Microorganisms use organic materials as energy sources. Consequently, large populations of microorganisms develop wherever and whenever the materials are available.

Microorganisms in the rhizosphere influence plants in many, usually helpful ways. Uptake of nutrients (e.g. phosphorus, potassium, molybdenum, manganese) is improved by microorganisms. This probably occurs because microorganisms produce water soluble compounds that help release some nutrients from soil minerals. Physiology and development of plants is faster when growth factors produced by microorganisms are present.

Some microorganisms in pasture soils inhibit nitrification (oxidation reaction that forms nitrate from ammonium). This means that pasture plants must depend more on ammonium for their nitrogen needs than on nitrate. This also means that losses of nitrogen through leaching (loss of nitrate) or denitrification (reduction reaction that uses nitric acid from nitrate to form nitrogen gas, which is lost) are minimal under pastures.

Legume-Rhizobium Symbiosis

One of the most important relationships between plants and microorganisms in pastures (and elsewhere), is the symbiosis between legumes and rhizobia bacteria. A symbiosis is a biological relationship between two or more organisms that benefits all of the organisms involved. In this case, the legume-rhizobia symbiosis benefits themselves plus all other organisms. Life as we know it would not be possible without this symbiosis that makes atmospheric nitrogen available to living organisms. Many

45

things affect the legume-rhizobia symbiosis, but here let's just look at the ones that we can influence to help it work better.

In this symbiosis the legume plant determines whether or not it will be nodulated by rhizobia. The plant releases certain organic materials from its roots which develop a rhizosphere that's favorable to rhizobia. Once the rhizobial population builds up, some bacteria infect the plant, and the plant forms a nodule around each bacterium. Within nodules the bacteria multiply and change to a form that is no longer free-living (bacteroid), but must depend on the plant for nutrients to support it. In return for nutrients and protection, bacteroids fix nitrogen, which becomes available to the plant.

For this symbiosis to function well, all required nutrients must be available in adequate amounts. Besides the nutrients needed by plants, cobalt must also be present because the bacteria require it. The nutrients specifically involved in nodule formation and nitrogen fixation are molybdenum, sulfur, copper, boron, iron, and cobalt. Any nutrient deficiency for the plants reduces their growth and vigor, and results in less nitrogen fixation.

Soil pH must be at the level required by rhizobia for nodulation and nitrogen fixation to occur. Rhizobia differ in their soil pH requirements. The rhizobia that infect alfalfa perform best at pH 6.5. The rhizobia that infect red and white clover tolerate lower soil pH levels, but do best at pH 5.5 to 6.0.

Light affects the symbiosis in at least two ways. The main effect involves photosynthesis and the subsequent transfer of photosynthesis products to root nodules. There is a direct relationship between light intensity, nodulation, and nitrogen fixation. Low-light conditions reduce the amount of carbohydrates available to nodules, and nitrogen fixation slows or stops.

Another effect of light on nitrogen fixation concerns

nodule development. Red light is required for nodules to form, and far-red light inhibits nodule development. In pasture swards when legumes are shaded by tall-growing grasses, red light is filtered out by the grasses. Light that reaches the legumes contains more far-red light than red light, and nodule development may stop. Maintaining a favorable red to far-red ratio for legume growth and nitrogen fixation is one of the reasons for keeping the sward short (less than 6 to 8 inches tall) through grazing management.

Some legumes tolerate shading more than others. If birdsfoot trefoil is only slightly shaded, its nodules drop off. Red clover and alfalfa tolerate shade more than trefoil. White clover tolerates shade better than trefoil, but not as well as alfalfa and red clover.

Adequate soil moisture is needed for nodule formation and nitrogen fixation. If conditions are too dry, fixation slows; if the stress is too severe, nodules drop off.

Too much water has similar effects. If soil is too wet, gas exchange can't occur between nodules and soil air. Low oxygen exchange results in reduced nitrogen fixation and eventual loss of nodules.

Defoliation can greatly affect rate and amount of nitrogen fixation. Legumes may shed all or most of their nodules every time they are grazed or cut. This happens because when leaves are removed, photosynthesis is drastically interrupted, and carbohydrates no longer move to nodules. On the contrary, carbohydrates move upward from storage to regrow leaves. So every time legumes are grazed, their nodules may fall off and decompose. Under favorable conditions, infection of the plants by rhizobia occurs again, nodules reform, and nitrogen fixation begins once more.

Earthworms

Every day active earthworms take in an amount of soil equal to their body weights. This activity makes a very important contribution to aeration and movement of soils.

Pasture soils contain three to four times more earthworms (about 1,200,000/acre) than tilled soils (400,000/acre). The weight of earthworms in a pasture soil may be twice as much per acre as the weight of livestock carried per acre on the surface! In a well developed Vermont pasture we found 1,109,000 and 1,776,000 earthworms/acre, with total weights of 2,318 and 3,454 lb/acre, under controlled grazing by cattle and sheep, respectively.

These numbers are directly reflected in the amounts of soil moved. For example, the total annual excrement of earthworms is about 21 tons/acre in old permanent pastures, compared to only 9 tons/acre on tilled land. This means that earthworms move at least 21 tons of soil per acre in permanent pastures each year! As pastures get older and grazing management favors plant growth, earthworm numbers and amount of soil moved increase.

Earthworms not only move soil, but also make some elements more available for plant growth. When earthworm excrement is compared to the top 6 inches of soil, the excrement contains five times more nitrate nitrogen, twice as much calcium, almost three times more magnesium, seven times more phosphorus, and 11 times more potassium than soil. The excrement also has a higher pH than surrounding soil.

Soil acidity generally is unfavorable to earthworms. Some earthworm families are sensitive to the high hydrogen ion concentration of acid soils, and others are sensitive to the low calcium availability of acid soils.

An adequate level of nitrogen in a pasture favors the

development of a larger earthworm population. This is probably due to more shading (which keeps the earthworms cool and moist) as a result of better plant growth with available nitrogen. Also earthworms have more food at their disposal in the form of fallen leaves and dead roots from the increased plant growth. A good level of available phosphorus is very beneficial to the development of earthworms.

The return of animal excrement to the soil increases the number and individual weights of earthworms present in a soil. This is extremely important in a pasture with regard to the breakdown and decomposition of animal manure. When a pasture environment is in good condition, it contains large numbers of earthworms and other organisms that rapidly break down and decompose the organic matter of manure, releasing the nutrients it contained into the cycle again.

Legumes such as white clover seem to have an especially beneficial relationship with earthworms. Earthworms feed on dead legume residue, and legumes gain from the improved soil fertility due to earthworm excrement.

Nematodes

Despite their poor image, nematodes are beneficial and important, due to their vast numbers in soils (We found 2,810 to 15,208 nematodes per 100 grams of soil under well developed pasture in Vermont). Activities of dominant nematodes result in rapid decay and incorporation of organic matter within the soil, and nutrient cycling. Some nematodes feed on bacteria, fungi, and soil protozoa, thereby assisting in the natural balance of other soil life forms. Destructive or pathogenic nematodes are absent or present in very low numbers of less than 1 percent of the nematode population in nontilled soils. When they are present in soils in good condition, pathogenic forms are kept in check by predatory nematodes.

PESTS

The presence of a large number of pests of any kind in a pasture is usually a symptom of poor management that interferes with the natural balanced order of the ecosystem. Problems arise when we do something to part of an environment, without having the slightest idea about the possible effects that our interference might have on the whole ecosystem.

Weeds

There really are no "weeds" in a properly managed pasture, except for plants such as thistles and poisonous plants that simply can't be grazed without harming the livestock. But we interfere with the pasture environment when we under- or overstock a large area and continuously graze it with our livestock, that aren't free to move on after they have grazed it once, as they would do under natural conditions.

The animals graze first and repeatedly plants that they like best, such as clovers, lush grasses, and dandelions. They leave plants such as smartweed and pigweed because the plants have a smell or taste that the animals dislike, or they are covered with short fuzzes that tickle the animals' mouths. Despite being a little less palatable, these so-called "weeds" have high feeding values in immature growth stages. If these "weeds" are not eaten along with the clovers and grasses, the "weeds" are given a competitive advantage. The "weeds" mature, set seed, and multiply, spreading throughout the pasture, progressively decreasing the amount of more desirable plants in the pasture.

In an overstocked continuously grazed pasture, animals graze all plants repeatedly, except thistles and other extremely obnoxious plants, every time the plants grow tall enough to be grasped by the animals. By early July such pastures appear to be producing nothing to sustain the animals. Under such grazing pressure many desirable plants assume an extremely low-growing form that allows them to escape being grazed. White clover, for example, will produce flowers 1 inch or lower from the soil surface, depending on whether they're being continuously grazed by cattle or sheep. Thistles and similar plants take over the pasture, since all of their competition is removed by the grazing.

Usually pastures are terrific messes by the time people begin using Voisin controlled grazing management. Almost any plant that protects itself from being grazed by having thorns, fuzzes, bad taste, bad smell, or poison can be eliminated simply by cutting them off. They may have to be cut off more than once, because some of them regrow from root reserves, but usually two or three cuttings will take care of them. Controlled grazing management favors desirable plants, which will become stronger and eventually compete successfully against even the most obnoxious weeds, if the weeds are cut repeatedly.

Careful use of systemic herbicides (e.g. glyphosate) to individually treat noxious weeds may eliminate them quicker than by cutting. Spraying large pasture areas with herbicides is not advisable, because herbicides affect all plants, not just the ones you want to get rid of. Herbicides also affect soil organisms in complex ways that may or may not be to your benefit. Besides, herbicides and other pesticides are hazardous to the health of all organisms, including you. First try all of the cultural practices that are applicable, to resolve weed problems. If you must resort to using herbicides, be extremely careful!

Poisonous plants have to be treated carefully. Don't

force your animals to graze down any area that contains a lot of plants that are poisonous to livestock. Normally animals know better than to eat such plants, but if you starve them into eating poisonous plants, they will do it. Remove poisonous plants from your pasture before you have your animals graze it down closely.

Insects

Insects of all kinds (locusts, grasshoppers, crickets, leafhoppers, aphids, weevils, grubs, termites) live in pastures and damage them to some extent. The damage results mainly from them eating leaves, boring into stems, sucking plant juices, introducing fungi and viruses, and eating roots.

In a balanced environment, the damage by insects isn't severe or important. Birds, diseases, parasites, other predators, and adverse weather all help to control insect populations.

Insect problems in pastures can be brought on by things that we do which change conditions of the pasture area. Overgrazed (i.e. continuously grazed) pastures, for example, provide ideal situations for breeding of grasshoppers and locusts. When woodlands are reduced, and ponds and marshes drained, bird populations are affected, and conditions may become favorable for development of larger insect populations. It has been estimated that 15 locusts or grasshoppers per square yard of a 40-acre pasture eat the equivalent of 1 ton of hay each day. So it certainly is worthwhile to do things (e.g. provide birdhouses) that encourage large numbers of insect-eating birds to live in or near pastures. Birds also help break up cattle dung and speed its decomposition, by digging in the dung for seeds and insect larvae.

Flies can cause serious problems to grazing livestock. Like other insect pests, however, flies decrease in number in properly managed pastures. This secondary benefit of good pasture management may result from more uniform distribution of manure, and its rapid breakdown and decomposition. Large bird populations also may help decrease fly problems by eating flies and their larvae.

Diseases

About 45 diseases are capable of attacking pasture plants in the United States. Grazing management that results in a mixed population of vigorously growing plants is probably the best control of these diseases, which tend to infect weak plants. Pastures should be grazed down closely in the fall to prevent the overwintering of disease organisms on leaves. This helps decrease reinfection of plants in spring.

Rabbits, Hares, and Rodents

Rabbits, hares, and various rodents (prairie dogs, woodchucks, gophers, mice, rats, ground squirrels) often seriously damage pastures and rangelands. Overgrazing creates conditions that favor development of high populations of these animals, and thereby contributes to increasing the damage done by them. Destruction of useful kinds of wildlife (wolves, coyotes, wild cats, owls, eagles, hawks, and many snakes) also allows rabbit, hare, and rodent populations to increase excessively. In California, for example, it was estimated that ground squirrels were eating as much forage each year as 160,000 cattle! Encouraging abundant diversified wildlife populations, and properly managing pastures are the best ways to minimize damage from these animals.

CHAPTER 2

SHADE FOR LIVESTOCK

Providing shade for grazing animals isn't a simple matter. If a pasture has too many trees production can be improved by thinning them to cover about 30 percent of the area, with as uniform a distribution as possible. This amount of trees provides adequate shade, and doesn't interfere with solar energy interception by pasture plants.

The usual pasture has two or three trees or groups of trees at various locations. Animals tend to graze away from the trees, but camp around them. Although this transfers nutrients to the immediate area around the trees from the rest of the pasture, and in the long term decreases pasture productivity, it's a small price to pay for having valuable trees on your land. The Earth needs all the trees that can be grown.

Establishing trees in a pasture to provide uniform shade is neither easy nor inexpensive. Besides the cost of buying and planting trees, each sapling must be protected from being eaten by grazing livestock. I've tried various ways of establishing trees on pasture, including spraying saplings with manure tea and using tree shelter metal mesh or plastic tubes (e.g. Tubex tree shelters from Trident Enterprises, PO Box 1774, Frederick, MD 21702), with and without electric fence. Manure tea seems to make the saplings more attractive to livestock. Without electric fence, animals rub on the tree shelter tubes until the tubes are knocked over, and then the saplings are eaten. The only way that I've found that works is to plant saplings in rows in tree shelter tubes, protected on both sides of each row with electric fence that allows livestock to just barely graze around the tubes without touching them. The rows should begin at paddock perimeter fences and extend only partly across paddocks, so animals can move around the rows of growing trees. Single-strand fence can be used to protect trees from cattle and horses, but multiple-strand fence is needed to protect them from goats and sheep.

If you decide to plant trees on pasture to provide shade, you should consider choosing tree species that allow you to double crop trees and pasture (see J.R. Smith *Tree Crops* in References. For example, black walnut trees produce extremely valuable lumber; honey locust trees grow quickly, fix nitrogen, and provide nutritious pods for animal feed in autumn.

Some dairy farmers let their cows out to graze in the morning, and keep them in the barn during the heat of the afternoon. This requires some feeding of the animals in the barn, which substitutes for pasture forage they otherwise would have eaten, as well as cleaning the barn and spreading manure -- all of which take time, labor, money, and energy. It also results in nutrient transfer from the pasture to the lanes and barnyard, and wastes the animals' energy in walking to the barn.

If you consider the conditions that existed in the vast treeless prairies where huge herds of grazing animals lived, it appears that livestock probably don't require shade, although they may prefer it when it's available.

REFERENCES

Bergersen, F.J. 1982. *Root Nodules of Legumes: Structures and Functions.* Res Studies Press, London.

Davidson, R.L. 1978. Root systems--the forgotten component of pastures. p. 86-94. In J.R. Wilson (ed) *Plant Relations in Pastures.* CSIRO, Melbourne, Australia.

Hardy, R.W.F. 1977. *A Treatise on Dinitrogen Fixation. II. Agronomy and Ecology.* John Wiley and Sons, Inc., NY.

Harper, J.L. 1978. Plant relations in pastures. p. 3-16. In J.R. Wilson (ed) *Plant Relations in Pastures.* CSIRO, Melbourne, Australia.

Harris, W. 1978. Defoliation as a determinant of the growth, persistence and composition of pasture. p. 67-85. In J.R. Wilson (ed) *Plant Relations in Pastures*. CSIRO, Melbourne, Australia.

Ludlow, M.M. 1978. Light relations of pasture plants. p. 35-49. In J.R. Wilson (ed) *Plant Relations in Pastures*. CSIRO, Melbourne, Australia.

McWilliam, J.R. 1978. Response of pasture plants to temperature. p. 17-34. In J.R. Wilson (ed) *Plant Relations in Pastures*. CSIRO, Melbourne, Australia.

Rhodes, I. and W.R. Stern. 1978. Competition for light. p. 175-189. In J.R. Wilson (ed) *Plant Relations in Pastures*. CSIRO, Melbourne, Australia.

Rovira, A.D. 1978. Microbiology of pasture soils and some effects of microorganisms on pasture plants. p. 95-110. In J.R. Wilson (ed) *Plant Relations in Pastures*. CSIRO, Melbourne, Australia.

Semple, A.T. 1970. *Grassland Improvement*. Leonard Hill Books, London. 400 p.

Smith, B., P.S. Leung, and G. Love. 1986. *Intensive Grazing Management*. Graziers Hui, Kamuela, Hawaii. 350 p.

Smith, J. R. 1929. *Tree Crops*. Harper & Row, NY. 408 p.

Turner, N.C. and J.E. Begg. 1978. Responses of pasture plants to water deficits. p. 50-66. In J.R. Wilson (ed) *Plant Relations in Pastures*. CSIRO, Melbourne, Australia.

Vincent, J.M. 1974. Root-nodule symbiosis with Rhizobium. p. 265-341. In A. Quispel (ed) *The Biology of Nitrogen Fixation*. North Holland Pub. Co., The Netherlands.

Voisin, A. 1959. *Grass Productivity*. Philosophical Library, NY. 353 p.

Voisin, A. 1960. *Better Grassland Sward*. Crosby Lockwood & Son Ltd., London. 341 p.

Watts, A.W. 1957. *The Way of Zen*. Pantheon Books, New York.

3

Pasture Nutrition

The morning glory that blooms for an hour
Differs not at heart from the giant pine,
Which lives for a thousand years.

Anonymous

As far as we know, plants require 16 elements to live and grow. These are: boron, calcium, carbon, chlorine, copper, hydrogen, iron, magnesium, manganese, molybdenum, nitrogen, oxygen, phosphorus, potassium, sulfur, and zinc. Notice that I've listed them in alphabetical order. Any other order would imply that some are more important than others, when actually all are equally important, but needed in different amounts. Plants take at least these 16 elements and, with the driving force of solar energy, integrate them into living life forms. If any one of them is absent, life is absent. All life therefore derives from the sun, and plants' ability to use solar energy to organize these elements.

So how can one element be more important than any other? We are the ones who define and categorize "things", as if the "things" were actually separate from all else. They are only separate in our minds, not in reality. Even our so-called holy books describe an unholistic creation of separate things, with man above it all, ruling nature. This reflects the mistaken perception that human beings have of the universe. This mistaken perception lies behind most of our troubles. It allows us to make war against other people, and to poison the Earth environment with pesticides and industrial wastes, because we feel that we are separate from it all, unrelated to the whole. Just try to live for one second without the Earth environment, and experience what a truly foolish notion separateness of things is.

This foolish notion is also why our pastures are such messes. Our usual management has always attacked certain aspects of pastures, such as in fertilization and renovation practices, but never has taken into account the whole pasture environment.

Everything is interrelated. We say that some elements, especially nitrogen, phosphorus, and potassium are most important, but this is because they are most important to us, since they are the ones needed in largest amounts. To plants, all are equally important, in balanced quantities.

All of the elements except carbon, hydrogen, nitrogen, and oxygen ultimately derive from the parent material giving rise to soil. Carbon, hydrogen, and oxygen come from the atmosphere and water. Nitrogen comes from the atmosphere through biological nitrogen fixation. The presence of these mineral nutrients and plants' success in obtaining them, mainly determine natural distributions of plants, and their ability to grow and survive in environments modified by humans.

One of the first steps in improving the management of your pasture is to test soils and/or plant tissues, and follow the recommendations that result from the tests. Soil and tissue tests aren't perfect, especially soil tests for pasture, since they were developed using tilled cropland soils. But the tests are the best tools we have for making sure that no major soil fertility or pH problems exist.

Follow the testing lab's instructions for taking and handling soil and plant-tissue samples. Test results can only be as good as the sample analyzed, so your most important part in testing is to take samples that represent the pasture soils and plants.

Since more than 85 percent of pasture plant roots are concentrated in the top 2 to 3 inches of soil, the top layer is most important in plant nutrition and in affecting sward composition and forage yield. The top layer is especially important in permanent pastures, because the soil isn't mixed with fertilizer and lime throughout a 6- to 8-inch depth as in tilled soils. As a result, fertilizer and lime applications to permanent pasture only affect this top soil layer. So to estimate fertilizer and lime needs, we should sample only the top 3 inches of permanent pasture soil.

The best way to take soil samples is with a soil probe (auger or tube). You can either buy a soil probe at a farm-supply store or through a farm-supply catalog, or sometimes borrow one from your county extension office. When using a tube soil probe, just push the point into the soil down to 3 inches, and pull it out straight up, without twisting the probe. If you twist a tube probe, it will break off. Augers are more expensive than tube probes, and aren't needed unless you have extremely stony or hard soils. When pulling a probe (especially augers) out of the soil, lift it out with your knees, never with your back.

If you don't have a soil probe, you can use a shovel. Just push the shovel in about 3 inches and pry the slit apart enough to get your hand in. Then scrape a handful of soil from the middle of the opening, starting from the bottom and scraping upward with your fingers.

If your pasture has obvious soil differences, such as hilltops, hillsides, and valleys, sample the different soils separately and have them analyzed as separate samples.

By this I don't mean that each hilltop should be a separate sample, but that all or several hilltops should sampled and combined together as one hilltop sample. The reason for this is that each soil type has different fertility characteristics, especially in pastures, where some areas may have received more manure and urine than others, because of the way that animals graze pastureland.

When sampling an area, move across it in a zigzag pattern, taking samples at the point of each zig and zag and midway between the two. Be especially careful to avoid sampling in or near obvious manure and urine patches. Collect 15 to 20 samples per 5 acres (composite sample), placing them in a clean plastic bucket that never has had any fertilizer in it. Then mix the samples together well, remove about a one-half pound subsample, and place it in a clean plastic bag. Use a permanent marker to label the bag. So that you know where the sample came from, either keep notes or draw a map of the pasture, indicating where the samples were taken. Empty out the bucket and begin taking the next composite sample. Keep soil subsamples in the shade until you have finished sampling, and refrigerated until they can be analyzed.

Follow the testing lab's fertilizer and liming recommendations, or calculate your own rates from the lab's analyses (see *The Farmer's Fertilizer Handbook* in References). No matter what grazing management is used, plants can't grow if they lack nutrients.

Now let's look at what happens to nutrients in a pasture environment, to understand how we might influence their availability, distribution, and flow to benefit pasture plants and grazing livestock.

NITROGEN

A grass-legume pasture gets its nitrogen almost entirely from the legumes through biological nitrogen fixation.

What does this mean? Well, to begin with, the amount ᵥ nitrogen needed each year in a pasture is simply amazing. Take for example a seasonal pasture yield of 5 tons of dry forage per acre. (We have measured 4 tons of dry forage per acre as an average amount produced in Vermont pastures between May 1 and October 1) Since this forage contains about 4 percent nitrogen, it means that 400 pounds of nitrogen per acre had to be available to the plants. If we also take into account nitrogen needed for root formation, and nitrogen losses by leaching and denitrification, the total amount of nitrogen available to the pasture environment had to be more than 555 pounds per acre during the growing season!

Nitrogen Cycle

Two main pathways exist in which nitrogen transfers from legumes to associated grasses: aboveground and underground.

Most nitrogen transfers aboveground, as nitrogen excreted by grazing animals. The amount of nitrogen removed from pastures by grazing animals is small. For example a wool fleece contains only 1.5 pounds of nitrogen, and a fat lamb contains about 4 pounds of nitrogen. These are tiny amounts compared to the total amount of nitrogen cycling in high-producing pastures. Most of the nitrogen (and all other elements, for that matter) remains in a well managed pasture, cycling continuously through the environment. So usually all that's needed in any pasture are adequate levels of nutrients to start with. Keep in mind that any forage, grain, and mineral supplements fed to pastured livestock, in effect fertilize the pasture with nutrients from the supplements that pass through in manure and urine. Generally there's no need for massive annual maintenance applications of any nutrients to well es.

mall amounts of nitrogen removed from
estock products, 70 percent of the rest of
ogen ingested in forage is excreted in urine and 30 percent in manure by grazing animals. Nitrogen in urine mainly exists as urea (70 percent) and amino acids; all are quickly converted to readily available ammonium and nitrate by soil microorganisms.

Livestock grazing a pasture containing only 30 percent white clover, apply nitrogen in urine patches at a rate of 225 lb of nitrogen per acre. Assuming no overlap of urine patches occurs, a pasture capable of carrying one animal unit (1000 lb liveweight) per acre during the grazing season, would receive an annual nitrogen application equivalent to 2,250 lb of ammonium sulfate per acre!

Put another way, 100 cows grazing 1-acre paddocks for 24 hours each, apply 43 lb of nitrogen, 10 lb of phosphate, and 35 lb of potash in their manure and urine per acre per day!

In contrast to nitrogen in urine, the nitrogen in manure isn't immediately available, because it exists in organic combinations that must first be broken down and mineralized by birds, insects, earthworms, and soil microorganisms.

Two other processes contribute to aboveground nitrogen transfer: leaching of nitrogenous compounds from plant shoots, and decay of leaves and petioles that have fallen from plants. In an ungrazed situation such as a hayfield, these are the only ways that nitrogen transfers aboveground to associated grasses. Much less nitrogen transfers in these ways, compared to that transferred in urine and manure.

Underground nitrogen transfer occurs through the leaking of nitrogenous compounds from legume nodules, and the sloughing off and decay of nodules and root tissue. Since white clover roots and nodules contain 1.5

and 6 percent nitrogen, respectively, a lot of nitrogen can be transferred to grasses underground, but not nearly as much as transfers aboveground with grazing animals

Unfortunately, grazing animals don't distribute manure and urine evenly over the pasture, and this is the main reason for losses from the nitrogen cycle. Animals tend to lie down or camp on certain spots within a pasture. The more irregular the pasture terrain, the more the animals congregate in their favorite spots. This effect can be seen especially in hilly country, where livestock graze on slopes and valleys during the day, but spend nights on hill crests. Anyone who has watched animals at all, knows that when they get up after lying down, the first thing they do is poop and pee. If they are always allowed to camp in their favorite areas, nutrients will be transferred from the rest of the pasture to those spots. Over time, excessive nutrient levels accumulate in those spots, while the rest of the pasture becomes deficient in nutrients.

The amount of nitrogen excreted not only varies over different parts of a paddock, but also varies within the patches of manure and urine that are deposited. For example, one urination of a beef cow affects a more or less circular area of 21 to 25 inches in diameter, or 2.4 to 3.4 square feet. Most of the nitrogen in a urine patch is concentrated near the center of the patch. Because of both the uneven distribution of urine throughout a pasture and within urine patches, nitrogen levels in the soil in livestock camps and within urine patches may get as high as 1,500 lb/acre! Even grasses with their large capacity to absorb inorganic nitrogen are unable to take up all the nitrogen present. This leaves some ammonium nitrogen available for conversion through nitrite to nitrate by soil microorganisms. Unfortunately, nitrate nitrogen is easily leached from soil, and, besides being lost to the pasture environment, this nitrate nitrogen pollutes surface and ground water.

Another form of nitrogen loss from urine patches occurs through denitrification, in which nitrogen gas is produced by microorganisms from nitrites and nitrates. Also, about 12 percent of the nitrogen present in a urine patch escapes as ammonia gas, giving urine its characteristic smell. The nitrogen gases return to the atmosphere, possibly to be fixed again in legume nodules and reenter the pasture environment.

Although nothing can be done to even out the distribution of nitrogen within urine patches, controlled grazing management helps to minimize the concentration of manure and urine in livestock camps. Dividing pastures into paddocks and consequently increasing stocking density within the paddock area, reduces the concentration of excreta in camp sites. As stocking density increases, less manure and urine are deposited in camp areas, possibly because animals spend less time at any one camp site, as they graze more evenly over a paddock.

Dividing pastures into paddocks doesn't eliminate the problem of nutrient transfer to camp sites, however, because animals soon find preferred camping spots within each paddock. Despite this tendency, the concentration of nutrients is less because they are only being concentrated within the smaller paddock area, and not from the entire pasture.

Try to divide pastures so that spots likely to be preferred for camps are separated from areas where just grazing occurs. For example, try to place hill crests in paddocks separate from slopes or level areas. Also, slopes of different aspect should be separated; that is, shady slope faces should be separate from sunny slope faces, which likely would be camp areas. But if this isn't possible, don't worry about it. Probably the biggest beneficial affect on excrement distribution comes from the increased stocking density per unit area that occurs when pastures are divided into paddocks.

Nitrogen And Grass-Clover Balance

Nitrogen applied either in fertilizer or animal excrement to a pasture sward suppresses clover growth, as a result of the increased competition from associated grasses. As discussed in Chapter 2, competition may occur for light, water, and nutrients, but usually involves more than one of these. This is because competition for a nutrient or for water has the secondary effect of resulting in different heights of the competing plants, which causes shading of lower growing ones and, therefore, competition for light. Under most pasture conditions, available nitrogen supply limits grass growth. So when nitrogen is applied, grasses use it to produce a large amount of topgrowth that shades associated clover plants.

Shading restricts root growth more than top growth. Shaded plants then have less capacity to use water and nutrient supplies. This restriction further limits top growth and light interception. The effect of nitrogen application in suppressing clover growth in a pasture, therefore, is due to the nitrogen causing an increase in both light and nutrient competition by associated grasses.

Applying nutrients other than nitrogen that previously limited clovers, usually gives a spectacular burst of clover growth. This is because in the first stages of a grass-legume sward's response to applied nutrients, lack of nitrogen still limits grass growth, and clover competes best for light, water, and nutrients. As soil nitrogen builds up by underground or aboveground transfer of nitrogen from vigorously growing clover, grasses no longer are limited and begin to compete more for light, water, and nutrients.

Although legumes have a high light requirement, it

doesn't seem to be for physiological reasons. It's probably due to the plants' growth habit, especially that of white clover. White clover's low-growing habit with a very dense, almost single layered leaf canopy isn't very efficient in intercepting light, compared to the more erect growth of grasses. The suppression of clovers by nitrogen application or buildup can be reduced somewhat by frequent, close grazing to lessen the advantage of grasses in competing for light.

Competition between grasses and legumes occurs for other nutrients besides nitrogen. Grasses usually have an advantage over legumes in obtaining nutrients, because grasses have a more fibrous root system. Legumes generally can't get enough nutrients until the needs of associated grasses have been satisfied. Fertilizer applications to pastures should therefore supply adequate levels of all nutrients other than nitrogen to meet the needs of both grasses and legumes.

PHOSPHORUS

Besides nitrogen, phosphorus is the other most commonly deficient nutrient. Phosphorus contents range between 0.35 and 0.45 percent in pasture plants growing on soils containing adequate amounts of phosphorus. So, a pasture producing 5 tons of dry forage per acre during a season must have about 42 lb of phosphorus available per acre.

As with nitrogen, very little phosphorus leaves the pasture environment in livestock. For example, cow milk contains only 25 percent of the phosphorus eaten in forage, and a fat lamb contains only about 0.4 pound of phosphorus. Although little phosphorus is removed from pastures by livestock, pasture forage should contain an adequate amount of phosphorus for balanced animal nutrition.

In contrast to most other plant nutrients, phosphorus usually is adsorbed by soils and becomes only slightly soluble, so that losses from pastures by leaching are very small. So once an adequate level of soil phosphorus is reached, only small maintenance applications are needed because most of the available phosphorus continues to cycle within the pasture.

Grasses compete more strongly than legumes for phosphorus and, as discussed above, their competitive ability increases when high levels of nitrogen are present. Competition for phosphorus by grasses is an important inefficiency in grass-legume pastures that depend only on nitrogen fixed by the legumes. This fixed nitrogen is not exactly cost-free, because large amounts of phosphorus have to be used in meeting the needs of grasses before legume needs can be satisfied.

Most phosphorus excreted by grazing livestock is in manure. Of this, up to 80 percent is present in an inorganic form immediately available to plants. The rest of the phosphorus in manure is present in organic combinations that must be broken down by birds, insects, earthworms, and microorganisms before it becomes available once again to plants. Like nitrogen, phosphorus accumulates in livestock camps, transferring in from the rest of the pasture area. The same precautions and management that spread nitrogen more uniformly over paddocks, also distributes phosphorus more evenly.

POTASSIUM

Luxury consumption doesn't just refer to affluent conspicuous consumers, but also to the gluttonous tendency of grasses to take up amounts of nutrients way in excess of their needs. This is especially true in the case of potassium. Because of luxury consumption, potassium content of grasses may rise to 5 percent of dry matter, although normal pasture forage contents range from 2.0 to 2.5 percent of dry matter. So, without luxury consumption of potassium, the annual requirement of a pasture producing 5 tons of dry forage per acre would be about 225 pounds of available potassium per acre.

The amount of potassium removed from a pasture in livestock products also is relatively small. For example, one milking cow removes only 9 pounds of potassium during the grazing season. Sooner or later though, this would deplete most of the potassium in a pasture, if more didn't continually become available from potassium-bearing soil materials or from applied potassium fertilizers.

Losses of potassium from the pasture environment can be very high, however, if manure and urine are distributed unevenly within the pasture. All excreted potassium is water-soluble; 90 percent of it is in urine, and the rest in manure.

As with nitrogen and phosphorus, grazing animals collect potassium from the whole pasture or paddock and concentrate most of it within urine patches near or in stock camps. These small areas may represent only 15 to 20 percent of the total pasture or paddock area. The end result is a general decrease of potassium over most of the pasture or paddock, and excessive amounts of potassium in camps.

Within a urine patch, potassium may be applied at a rate of over 1,200 lb/acre! Even with luxury consumption by grasses, this amount can't all be taken up by plants and some is lost by leaching. An amount equivalent to 150 pounds of potassium per acre may be lost from one urine patch. This has serious consequences in that it depletes the pasture of potassium and may pollute ground water.

Plant growth responses within urine patches indicate helpful things about a pasture's nutrient status. These responses usually become apparent about 2 weeks after urine is deposited. First the grasses become darker green and grow more vigorously, as they are stimulated by nitrogen in the urine; this happens at the clover's expense, remember. Three to 4 months after the urine was deposited, the grasses use up the urine nitrogen. Once the nitrogen is gone, clovers respond to any potassium remaining in the urine patch, and reach their greatest growth within the patch at this time. As the potassium once again becomes limiting, clover content drops to its previous low level. So clover dominance in urine patches indicates potassium deficiency in a pasture or paddock.

CALCIUM

Calcium is required by plants in amounts nearly as large as those of nitrogen, phosphorus, and potassium. In fact, some plants contain more calcium than potassium. Calcium is important not only as a plant nutrient, but also in its relationship to soil acidity and soil structure.

Acid soils may contain too much aluminum, iron, and manganese, and too little calcium and magnesium. Applying calcium in liming materials corrects soil acidity and neutralizes byproducts of organic decomposition. Soils generally contain relatively large amounts of calcium though, and, except for acid soils with pH levels below 6.0, usually enough calcium exists to meet the needs of growing plants.

Although calcium is related to soil acidity, a soil's pH doesn't necessarily indicate the amount of calcium that's available; a separate test is needed to determine how much calcium is in a soil. Calcium content of a soil mainly depends on the parent material from which the soil developed. For example, calcium content may range from 0.5 percent in coastal plain sands, to more than 5 percent in soils of dry regions or in soils that developed from limestone, chalk, or marl.

69

Calcium improves soil structure by aggregating colloidal clay and humus particles, which makes soil more granular. Granular soil allows air and water to enter it easily, and this provides a favorable environment for plant roots and soil organisms to live and grow.

Calcium is involved in all sorts of processes in plants, including root and leaf development, nodulation and nitrogen fixation, cell elongation and division, protein synthesis, and water uptake.

Despite the many essential functions of calcium in pastures, calcium deficiency is rarely observed in the plants. This probably is because other problems, such as aluminum toxicity or phosphorus deficiency develop first on poor soils, and limit plant growth before calcium deficiency symptoms can be expressed. Grazing animals require large amounts of calcium, however, and may suffer from calcium deficiency if the plants they are eating contain low levels of calcium, even though the levels aren't low enough to limit plant growth.

If you need to apply calcium, either to raise soil pH or to increase the amount of calcium available in the soil, probably the best material to use is finely ground limestone (calcium carbonate). If other nutrients are needed besides calcium, you can apply ordinary superphosphate (phosphorus, calcium, sulfur), dolomite (magnesium, calcium), or gypsum (sulfur, calcium).

MAGNESIUM

Magnesium is intimately involved in photosynthesis, and chlorophyll contains about 2.7 percent magnesium. So if there's a shortage of magnesium for plants, photosynthesis will be limited and so will forage production. Because it's so important in chlorophyll formation, the first symptom of magnesium deficiency is loss of healthy green color between leaf veins. The color gradually changes to yellow and then to reddish purple, as magnesium deficiency becomes more severe.

Besides being essential for plant growth, magnesium exists in every cell of animals and humans. The most common problem of magnesium deficiency in animals is called grass tetany, grass staggers, or hypomagnesaemia.

Magnesium content of soils varies a lot, depending on the parent material from which soils develop. Magnesium is most likely to be deficient for plants on sandy soils that have been limed with calcitic limestone and have received high rates of nitrogen and potassium fertilizers. In Vermont it's recommended that soils contain a 2:1 ratio of potash to magnesium, at levels of at least 300 lb of potash and 150 lb of magnesium per acre. Magnesium fertilizers include dolomitic limestone, magnesium oxide, magnesium sulfate (Epson salt), and potassium-magnesium-sulfate (sulpomag).

SULFUR

The cycle of sulfur closely resembles that of nitrogen. Most sulfur in soils is present in an organic form, and only becomes available to plants after it becomes mineralized through microbial action. Since grasses are very competitive and may use 95 percent of any sulfur mineralized from soil organic matter, legumes must depend on other sources of sulfur.

The amount of nitrogen fixed by legumes in a given environment determines the amount of sulfur that they need from sources other than soil organic matter. If sulfur fertilizer isn't applied, the only other source of sulfur is from the atmosphere in rain and snow. The amount of sulfur required by legumes equals one twelfth of the nitrogen fixed. So white clover fixing 250 pounds of nitrogen per acre during a season, needs 21 pounds of sulfur per acre! The amount of sulfur needed in a flourishing pasture is higher than what usually comes from the atmosphere in precipitation. Even acid rain may apply less than 10 pounds of sulfur per acre each year in rural areas away from industrial centers.

Until recently, phosphorus fertilizer generally used (ordinary superphosphate) contained about 12 percent sulfur. In concentrating phosphorus to about 45 percent of the fertilizer material (triple superphosphate), freight costs were lowered, but sulfur was eliminated. Consequently, in areas where only triple superphosphate is used, no sulfur is being applied to cropland. Of course, permanent pastures in the United States usually haven't been considered worthy of fertilization, so they weren't intentionally fertilized with any nutrients, including sulfur. Sometimes manure that contained sulfur was spread on permanent pastures, but this mainly benefited pasture grasses. So sulfur deficiency may be limiting plant growth in pastures, especially permanent pastures.

Many soil- or tissue-testing labs don't analyze for sulfur, because the analysis is more difficult and time-consuming than for other elements. One way to find out if sulfur levels are too low in your pasture is to spread 100 pounds (to be certain enough is applied) of sulfur per acre as gypsum (calcium sulfate) evenly over a few measured areas of your pasture. For example, measure off 5 by 20 feet (100 square feet) areas in different locations. Since gypsum contains about 15 percent sulfur, you will need to spread 1.5 pounds of gypsum over each 5- by 20-foot area, to apply 100 pounds per acre of sulfur to that area. Assuming that other nutrients are present in adequate amounts (and you can determine this by soil testing and following the recommendations), within 1 to 2 months plants should begin flourishing in the sulfur-fertilized areas. If this occurs, your pasture needs sulfur, and you should apply at least 25 pounds of sulfur per acre each year.

Probably the best way to apply sulfur is along with

phosphorus in ordinary superphosphate, or along with potassium and magnesium in potassium-magnesium-sulfate. If you don't need to apply other nutrients besides sulfur, use gypsum. Don't apply elemental sulfur (yellow, rotten-egg smelling stuff), because sulfuric acid results from the microbial transformation of elemental sulfur to sulfate that occurs in soil, and this lowers soil pH.

BORON

Boron was first shown to be a deficient nutrient in Vermont. Alfalfa and the clovers are very sensitive to low levels of boron in soils. A symptom of boron deficiency is shortened, rosette-shaped plants. Leaves turn yellow and look as if they have been damaged by drought. Boron deficiency symptoms are most likely to develop during dry spells and after applying large amounts of lime. For this reason, lime applications in Vermont are limited to 2 tons per acre per year.

No more than 2 pounds of boron per acre should be applied every 3 years to grass-clover pastures. Although the legumes may need more boron than this, higher rates might kill associated grasses. If boron is deficient in your area, follow the recommendation of your soil testing laboratory in amounts and times of application. Be very careful in applying boron; too much can easily kill grasses. It's best to test the boron application in 5- by 20-foot test areas or in a strip across the pasture, to make certain that the rate you apply won't kill grasses, before applying it to the entire pasture.

MOLYBDENUM

Research in Vermont showed that molybdenum may be nearly deficient in many soils of the state. We have gotten positive responses from applying molybdenum in combination with phosphorus and lime, on certain soils in which phosphorus and lime application alone had not improved plant growth very much. If you have paid attention to all other nutrients and your pasture plants still don't grow well after 2 or 3 years, molybdenum may be lacking. It is involved in nitrogen fixation, so if it's deficient, little nitrogen will be fixed.

Molybdenum is more available in alkaline soils and in soils having high organic matter contents. Liming soils makes molybdenum more available; in fact, response to liming may be due to better molybdenum availability in some soils.

Be certain that your soils actually need molybdenum before applying any, by having soils and plant tissues analyzed. Try liming the soil to raise the pH and make molybdenum more available before applying any molybdenum. Get a second opinion on any recommendation that involves applying molybdenum. Be extremely careful if you apply molybdenum! No more than 0.18 pound (80 grams) should be applied per acre, because it's easy to reach toxic levels of molybdenum. That amount can't be spread uniformly unless it's contained in a larger quantity of another fertilizer, such as ordinary superphosphate. As with sulfur and boron, it's best to apply molybdenum in test areas to see what happens before applying it to the rest of the pasture.

COPPER, CHLORINE, IRON, ZINC, AND MANGANESE

These nutrients generally seem to be present in adequate amounts in pasture soils, at least in the Northeast. If your pasture plants don't grow well after you have attended to all of the other elements, and you are using proper grazing management long enough (2 to 3 years) for soil organisms to develop to adequate levels, there is a slight possibility that one of these may be deficient.

Be careful that you don't apply these or any other nutrient unnecessarily. The best way to determine if you need any nutrient, including these, is by having soils and plant tissues analyzed and recommendations made by a lab that doesn't sell fertilizers of any kind. It's always best to get a second opinion on any fertilizer recommendation that involves applying these elements. Test any application of these elements in small areas before applying them to the rest of your pasture. If you apply too much of any of these or other nutrients such as molybdenum or boron to pastures, it's too late to avoid serious problems.

SOIL ACIDITY

Technically, soil pH is a measure of hydrogen ion activity in soil solution. Practically, it's an indicator of one aspect of soil that influences almost everything else in soil.

It's hard to determine the relationship between composition of a pasture sward and soil pH, because pH is intimately related to so many things in soil, and because it varies a lot in different layers of pasture soils. The pH value you get in a soil test is an average measurement of acidity over soil sample depth, but pH of the top 1 to 1.5 inch of pasture soil is known to be higher than that of the 4-inch layer underneath. Probably quite different nutritional and soil organism relationships exist in the top layer, compared to the layer beneath it. Also, soil of a permanent pasture usually has a lower pH than the same soil type under cultivation. The pH layers and lower pH of pasture soils need more research and accommodation by soil testing labs so that the information can be applied.

Soil pH directly or indirectly affects soil organisms, all plant nutrients, and plant growth. Plants that are well adapted to prevailing conditions grow best and make up the largest part of pasture swards.

75

As soil pH decreases below 5.5 to 5.8, bacteria become less active. This directly influences rates of nitrogen fixation and transformations, organic matter decomposition, and the sulfur cycle. The result is that nitrogen and sulfur are less available to plants. Calcium, potassium, magnesium, molybdenum, and phosphorus also are less available to plants in soils with pH lower than about 6.0. Boron is less available in soils when pH is below 5.0. Cobalt (required by microorganisms, but apparently not by plants), copper, iron, manganese, and zinc all become more available as soil pH decreases below 6.5. Aluminum, which is toxic to most plants, is much more available and influential below pH 5.5. In very acid soils aluminum and manganese can be present in toxic levels.

As soil pH increases, the reverse of all of the above happens. So it's clear that soil pH doesn't exist in a world by itself: every change in pH brings about many changes in other aspects of the soil.

The question then comes up of when is it necessary to lime a permanent pasture? Because of the pH differences in soil layers, and because certain plant species are particularly sensitive to changes in pH, it may be better to be guided by both the pasture plants that are present and a soil test in deciding whether to apply lime or not. For example, if there is a lot of sheep sorrel (*Rumex acetosella*) in a pasture, it definitely needs liming. But if the pasture is growing well and the plant species are desirable ones (this assumes proper grazing management), liming may not be needed. Generally it's best for pasture plant growth and productivity if soil pH is maintained in the range of 5.5 to 6.5.

Avoid spreading too much lime at any one time: only spread about 2 tons per acre. Then wait 2 to 3 years for the effects of that application to become evident before deciding if more is needed.

REFERENCES

Cramer, C., G. DeVault, M. Brusko, F. Zahradnik, and L.J. Ayers. 1985. *The Farmer's Fertilizer Handbook*. Regenerative Agriculture Assoc., Emmaus, Pennsylvania.

Follett, R.H., L.S. Murphy, and R.L. Donahue. 1981. *Fertilizers and Soil Amendments*. Prentice-Hall, Inc., Englewood Cliffs, New Jersey.

Jones, U.S. 1982. *Fertilizers and Soil Fertility*. Reston Pub. Co., Reston, Virginia.

Scott, W.R. 1973. Pasture plant nutrition and nutrient cycling. p. 159-178. In R.H.M. Langer (ed) *Pastures and Pasture Plants*. A.H. & A.W. Reed, Wellington, New Zealand.

Semple, A.T. 1970. *Grassland Improvement*. Leonard Hill Books, London. 400 p.

Smith, B., P. Leung, and G. Love. 1986. *Intensive Grazing Management*. The Graziers Hui, Kamuela, Hawaii. 350 p.

Vallis, I. 1978. Nitrogen relationships in grass-legume mixtures. p. 190-201. In J.R. Wilson (ed) *Plant Relations in Pastures*. CSIRO, Melbourne, Australia.

Voisin, A. 1959. *Grass Productivity*. Philosophical Library, Inc., New York. 353 p.

Voisin, A. 1960. *Better Grassland Sward*. Crosby Lockwood & Son Ltd., London. 341 p.

CHAPTER 3

Watts, A. 1972. *The Book: On the Taboo Against Knowing Who You Are.* Vintage Books, New York. 151 p.

Grazing Animals: Effects On Pastures & Vice Versa

The wild geese do not intend
To cast their reflection;
The water has no mind
To receive their image.

Zenrin

Grazing animals can make or break a pasture by causing very rapid and large changes in plant productivity and botanical composition of the pasture sward. These changes result from removal of plant leaves (defoliation), excretion of manure and urine by grazing animals, treading action of animals' hooves, and seeds dispersed by animals. Let's look at each of these separately to understand how they influence a pasture.

DEFOLIATION

Defoliation probably is the most important effect that grazing animals have on pasture. We saw in Chapter 2 how important light relationships are in a pasture sward in determining which plants will be dominant and which will be present in different numbers. Defoliation drastically decreases leaf area of plants that are grazed. This reduction of a plant's leaf area affects its carbohydrate use and storage, tiller and stolon development, and leaf and root growth. Defoliation also changes a plant's microenvironment of light intensity and soil temperature and moisture.

Plants differ in their physiological and growth-form responses to defoliation, and in their growth rhythms during the season. Because of these differences, grazing can change the relative abundance of the various plants in a pasture, depending on which plant species are best adapted to weather conditions after grazing. So dominant species and the composition of a sward can change several times during a season, as weather and light conditions change and are modified by grazing. If we made a time-lapse movie of these botanical changes, we would see them as waves passing through the sward, reflecting environmental and grazing influences.

When all or most of the leaves are grazed from a plant, root growth stops and the plant uses its carbohydrate reserves in roots and stubble for the energy it needs to grow new leaves and stems. The plant draws on reserves until enough leaf surface develops for photosynthesis to be adequate to meet energy needs for further leaf and root growth. Grasses depend on reserves for 2 to 7 days after grazing. This can take longer in other plants such as alfalfa, which needs about 21 days to form enough leaf surface for photosynthesis to provide adequate energy for subsequent growth.

Plant response and survival in a pasture depends greatly on use of reserves for regrowth, and the weather after grazing. Low temperatures and/or dry conditions slow regrowth of defoliated plants for long periods, during which the plants continue to draw on carbohydrate reserves. Continuous grazing, by repeatedly removing leaves and reducing reserves during and after adverse weather, decreases survival and regrowth of stressed plants. Development of cold hardiness, especially for alfalfa, may be lessened either by complete defoliation during the 6-week hardening period of September to mid-October, or by having leaves grazed off too many times during the season.

Grazing all or most leaves from a plant drastically affects its ability to take up nutrients and water from soil. If root growth stops or slows because of defoliation or any other reason, uptake of water and nutrients from soil is directly limited. When leaves are removed in grazing, flow of water from the soil through the plant to the atmosphere (evapotranspiration) decreases, and this also reduces uptake of nutrients. Removal of leaves also decreases the amount of carbohydrates being produced in photosynthesis and their movement to roots. In fact, carbohydrates may begin to flow entirely from roots to shoots as the plant regrows. During these times water and nutrient uptake decrease and the plant may be stressed even more.

Yin and Yang

Like everything else, defoliation of pasture plants has two sides to it. Reasonable amounts of defoliation by grazing causes more branching and lower growth forms of plants. These result in a tighter sod, more covering of soil against erosion, and higher yields of more nutritious forage. Also, plants having more branches or stolons, and less shading of branches or stolons, tend to have more flowers and produce more seeds. This tendency may be very important in maintaining a high percentage of red and white clovers and birdsfoot trefoil in intensively grazed pastures.

Another aspect of this other side of defoliation, is that the removal of nutrients from plants by grazing their leaves and stems, results in movement of nutrients from the soil root zone to plant tops. In this way, defoliation and digestion by grazing animals is an essential part of rapid cycling, return, and transfer of nutrients in a pasture environment. Uneaten plants or plant parts, in contrast, slow nutrient cycling because nutrients are unavailable until the plant material breaks down through weathering in the air or biological decomposition on the soil surface. Controlled high stocking densities reduce the amount of uneaten plant material, and help break down uneaten material so that it decomposes faster, thereby quickening nutrient cycling.

SELECTIVE GRAZING

Animals don't graze uniformly. That's why it's so difficult to have animals graze high to favor some plants. Animals select certain parts of a plant or particular plant species while ignoring others, if allowed to do so. This results in uneven grazing of pastures, which stresses individual plants differently in the sward, because of the variable vertical distribution of leaf, stem, floral, and growing tissues in the plants. Even when animals graze down to a relatively uniform height from the soil surface, some plants are still favored over others because of this variable vertical distribution.

Given the chance, all grazing animals are selective in their diet. Within a pasture, grazing animals generally have many choices of what to eat from a range of plant species with varying proportions of leaf parts, stem, flowers, and seed. Sheep and cattle tend to select leaf parts in preference to stem, and young leaves before old leaves, especially when pasture forage has become too mature. The plant material that animals select usually is higher in protein, phosphorus, soluble carbohydrates, digestibility, and energy, and lower in lignin and structural carbohydrates (fiber), compared to the total forage present in a pasture.

Animals know instinctively what they like and usually need. It seems that their selection of one plant species or plant part in preference to others depends on what we call palatability, accessibility, and availability.

Some plants or plant parts taste or smell badly or have picky structures that irritate animals' mouths. These kinds of things decrease palatability of plants and discourage animals from eating them. Some plants or plant parts aren't accessible to animals. Their mouth formations may not allow them to eat a plant, a plant may grow too low to the ground, or a plant part may be protected by thorns.

As pasture forage availability decreases, selectivity also decreases; forage that wasn't acceptable before, now will be eaten. Animals adjust, so that, while continuing to look for preferred plants or plant parts that are less accessible, they eat more plants or parts that they had previously rejected. In this stage animals take longer to graze because they spend more time searching for forage they prefer. They also take more bites, because their bites become smaller with decreasing forage availability.

As stocking density increases, differences in relative acceptability among plants or plant parts practically disappear. Animals will reduce intake, though, when left with greatly disliked plants or plant parts to eat. So be careful with high producing livestock, especially in early stages of pasture improvement.

Animal grazing selectivity is exactly the same thing that happens when children have a choice between eating lettuce or spinach: almost invariably they eat the lettuce and leave the spinach. If, however, they would be given the choice of eating both lettuce and spinach or starving, they would eat both. A grazing management method that gives animals the choice of eating all available pasture forage or starving, puts you in control of the situation and allows you to influence the botanical composition so that it benefits animals and pasture. You will find that most animals will choose to eat the forage before going hungry.

For all of the above reasons, different botanical compositions of pasture swards result under different grazing management practices. Close, continuous grazing with a high stocking density leads to a sward containing species that tend to be low-growing, with rhizomatous, stoloniferous, or basal rosette growth habits. Extensive continuous grazing with a low stocking density results in a sward rich in plant species, but these tend to be mostly tall-growers such as milkweed, goldenrod, burdock, and thistles. Rotational grazing with a low level of stocking density and forage use also results in a sward dominated by tall-growing plants. Controlled rotational grazing with a high level of stocking density and forage consumption usually results in a simpler pasture mixture of one or two legume species, several grass species, and two or three forbs (dandelion, chicory, plantain), which complement each other and are well adapted to the management.

If water and nutrients are not limiting, changes in a pasture sward caused by grazing mainly reflect grazing frequency and intensity. The changes result from defoliation, depending on whether and for how long light reaches levels in the canopy where low-growing plants (e.g. white clover) have their leaves. Arrangement of leaves and stems in a sward not only affects light penetration into the plant canopy and selective grazing, but also influences grazing time and forage intake.

GRAZING BEHAVIOR AND FORAGE INTAKE

A lot of research has been done on mechanical harvesting

and storage of livestock feed, but very little on how animals harvest their own forage while grazing pastures. If we understood more about how animals harvest their forage, we might be able to help them to eat more, and we could get better forage and livestock yields from our pastures. Although most existing information on grazing behavior is for cattle, a look at how cattle graze also can indicate how other kinds of livestock might graze.

Grazing time involves the entire process of moving around searching for food and then browsing (actually eating) the forage that is found. Grazing time in cattle almost never lasts more than 8 hours per 24-hour period. Browsing itself lasts less than 5 of the 8 grazing hours per day. This is an extremely important point: grazing time is the same regardless of pasture quality or amount of forage available. During the 8-hour grazing times, cows make all the effort that they can, and they can't graze any longer without resting. It is understandable why a cow would be tired after working for 8 hours to harvest forage, if you consider that a cow's mouth is only about 3 inches wide and she needs to eat about 150 pounds of green forage each day. Obviously, if we can help animals to harvest the forage they need, while using as little energy as possible, livestock productivity will improve.

About 60 percent of cattle grazing occurs in daylight, and 40 percent occurs at night. (Little if any sheep grazing occurs at night.) As air temperatures rise, the proportion of grazing done at night increases. Grazing animals walk as much as 2.5 miles per day, depending on forage availability. Eighty percent of the walking occurs during daylight, which corresponds to the fact that most grazing activity happens during the daytime. Probably grazing at night is more efficient, because animals are pestered less by flies and other insects at night, and consequently walk around less.

While grazing, cows move forward swinging their

heads from side to side within an arc of 60 to 90 degrees, and taking 30 to 90 bites per minute, if the forage is the right length. Forage length has an important effect on a cow's rhythm of eating. If a cow is grazing very long forage (10 to 14 inches), she either eats the upper 2.5- to 3-inch layer, or she tears off a mouthful about 12 inches long. If she takes in the long forage, she can't swallow such a large, long mass without chewing it first, and chewing it requires about 30 seconds per mouthful. By comparison, a cow grazing forage that is only about 6 inches tall, can swallow 30 mouthfuls in 30 seconds. The best forage height for sheep and goat grazing is about 4 inches tall. So animals grazing short forage can eat more during a day than when they graze long forage.

The amount of time that cows are capable of grazing and the number of jaw movements per day they can make apparently are inherited characteristics. In one study for example, a heifer made 29,150 jaw movements per day, compared to only 18,781 jaw movements in another heifer. The first heifer made 55 percent more jaw movements and grazed 46 percent longer than the second heifer. Animals that are capable of longer grazing times and more jaw movements can eat more than other animals that are limited in the effort that they can make. Selective breeding of animals for efficient use of pasture forage could result in less protein and energy supplements needed for animals grazing well managed pastures.

Cows ruminate about 8 to 10 hours per 24-hour period. Part of the ruminating occurs while cows are lying down, and part while they are standing up. Cows lie down for about 12 hours a day, and this total resting time usually is divided into nine unequal periods of 1 to 6 hours.

GRAZING HABITS

On top of all this, different kinds of animals graze

differently. If permitted, sheep and goats graze more selectively than cattle, because the large jaws and grazing action of cattle don't allow them to select as precisely as sheep and goats can. Also, milking goats have higher nutritional needs than milking cattle or sheep, and this causes goats to graze more selectively.

Since they only have front teeth on their lower jaws, ruminants grasp plants with their tongues and/or pinch forage between their lower teeth and upper mouth pads as they sweep their heads back and forth. Because of the position of their teeth and muscles of the lower jaw, cattle can't graze closer than 1/2 inch from the soil surface. Cattle can eat relatively mature forage, as long as the plants are not coarse or prickly.

Sheep and goats, in contrast, can cut plants off at the soil surface, taking away plant parts needed for regrowth. Uncontrolled sheep and goats may even tear up entire plants on overgrazed pasture. Although sheep and goats require very close management to prevent them from damaging pastures, a well managed sheep or goat flock can improve a pasture quicker and better than cattle can. This is because sheep and goats will eat almost everything under high stocking density, including young shoots of bushes, and don't have as much of a problem of avoiding forage around their manure droppings, as cattle do around cow pies.

Horses also can grip plants and cut them off closer to the ground than cattle can. Horses graze very selectively, making it difficult to get a good botanical composition in a pasture grazed only by horses. So it is best to follow horses with sheep or cattle, or graze them all together, to clean up what the horses leave. Otherwise you should mow each paddock after every grazing to cut down what the horses didn't eat, to keep the sward in good condition for subsequent grazing.

All livestock can be put out to pasture. All can be

managed in similar ways, as long as characteristics of that kind of livestock are taken into consideration. Pigs, for example, usually must have their snouts ringed before allowing them out. Otherwise they may dig up the soil and ruin the pasture, especially if you try to get them to eat everything in each paddock. If pigs are ringed and their grazing is controlled, a good quality botanical composition results. Geese and ducks tend to tear plants when eating, and their grazing favors development of plantain, sorrels, and smartweeds, because they prefer grasses to broadleaf plants. If confined too long in one area, chickens eat and dig up all of the plants that are present.

TREADING/TRAMPLING

Treading or trampling by animals' hooves is another aspect of grazing that influences and changes pasture botanical composition, because of its direct effects on plants and indirectly through its effects on soil and soil organisms. All livestock can cause treading damage, particularly when grazing is not controlled. The potential seriousness of this can be appreciated if you consider that an animal makes 8,000 to 10,000 hoof prints per day. If each hoof print covers 14 square inches as in cattle, the total area stepped on is 0.02 acre per day per animal! Horse hooves, especially when shod, may do a lot of damage to soil and to plants that can't withstand treading.

The affect of treading directly on pasture plants depends only slightly on soil type and fertility and plant height, but is strongly influenced by plant species and soil moisture. Plant species differ in their resistance to treading. Orchardgrass, perennial ryegrass, and white clover are very resistant to treading, compared to birdsfoot trefoil, red clover, and timothy, which are more sensitive. As stocking density increases, plants sensitive to treading decrease in number, and botanical composition shifts to plants that can resist the amount of treading that occurs.

Treading indirectly affects pasture plants by compacting (increase in density) and puddling (reduction in air space) the soil, especially if the soil is moist and fine-textured. Under moist soil conditions compaction and puddling result in increased mechanical resistance to plant root extension through the soil, thereby limiting water and nutrient uptake. They also reduce soil aeration, which slows root growth and metabolism, and limits nitrogen fixation. Temperature relationships also change when soils are compacted or puddled, and are reflected in above-ground growth in one way or another. Moisture availability changes in ways that may be favorable or unfavorable. Soil compaction and puddling decrease water infiltration and increase runoff, which leaves less water available to plants and erodes the soil. When soils are wet, plants and soil can be severely damaged by deep treading or poaching. Any of these things may depress plant growth, depending on how well the plants can adapt to the changed soil conditions.

Depending on animal species, treading effects on soil differ. On a Vermont fine sandy loam, for example, sheep grazing compacted the soil less than Holstein heifers, under a moderate stocking density of 80 animal units/acre/24 hours. Apparently because sheep weigh less per animal than cattle, sheep exert less pressure per square inch on soil.

Since plants, animals, and soils evolved together, treading has always been a part of their experience. Pasture plants, especially grasses, need to be grazed and probably treaded or trampled to avoid clumped dead growth that would shade growing points and hold nutrients from recycling. Under natural conditions, positive aspects of treading or trampling must have been more pronounced than negative effects. Otherwise, the huge grassland expanses of the world would not have developed under grazing by the enormous number of animals that lived in those regions. Fencing, domestic livestock, and poor grazing management changed all that, and detrimental effects of treading on plants and soil became evident.

Under controlled grazing management livestock are concentrated and limited in the amount of time they spend in a paddock. This in effect simulates natural grazing with its quick, concentrated grazing, recovery, and more positive treading effects on pasture plants and soils. For example, when dairy cows graze each paddock for 12 hours and rotate through the paddocks six times during a season, they only spend 3 days total time in each paddock per year. That leaves 362 days for plants, soil, and soil organisms to recover from grazing and treading, and make use of the nutrients that the cows left behind. This differs greatly from the effects of uncontrolled continuous grazing where livestock continually affect plants, soil, and soil organisms for several months each year.

(In a brittle environment, controlled trampling or herd effect can be beneficial in breaking soil crust, to improve water infiltration and seedling establishment. It also breaks down uneaten plant material, enabling decomposition to proceed more rapidly. See Allan Savory's *Holistic Resource Management* in References for information.)

EXCREMENT

Cattle poop 11 to 12 times and pee 8 to 11 times each day. When combined with grazing behavior, cattle and horse excretion especially can badly affect a pasture. When grazing is properly managed, manure and urine greatly benefit a pasture by returning nutrients, increasing soil organic matter, and favoring the development of earthworms and other soil life. Let's look at each part of excrement separately to understand the effects.

Manure

The daily amount of poop produced per cow can weigh 50 pounds or more. Horses are close behind with 40 pounds per animal per day. This means that during a 180-day grazing season, 9000 pounds of manure are deposited per cow, and 7200 pounds per horse! If each cow pie measures 10 inches in diameter, for example, the total area covered by one cow (assuming no overlap of pies) would be 7 square feet per day, or 1,260 square feet during the grazing season! Liquid manure from cows grazing lush pasture doesn't affect plants very much because it spreads thinly and decomposes quickly. But horse turds and cow pies from cows receiving grain concentrate supplements are drier, and cause major changes by blocking sunlight and killing most plants that are directly under the manure. The area in which the plants were killed can then be invaded by surrounding plants or from seeds that were in the soil or manure. Sheep pelletize their manure, so it's spread uniformly and breaks down rapidly when their grazing is managed properly.

When a cow pie hits the ground, it immediately has a "zone of repugnance" around it that measures about 25 feet in diameter! Consequently, at low stocking densities a lot of forage can be rejected around cow pies. High stocking densities can decrease the zone of repugnance down to the manure itself, but it isn't a good practice to force animals to eat right up to their manure because of the parasites it contains, and because dry matter intake probably would be decreased.

Horses deposit their manure in the same place and don't eat in that area. This causes forage in that area to become much lower in quality, unless it is mowed or grazed by other animals. Horse grazing and excreting habits especially, reflect their instinctive way of avoiding parasites present in their manure, so they shouldn't be forced to eat plants growing among their manure deposits. It is far better to graze that forage with other animals or mow and/or harrow it the same day that the horses are removed from the paddock.

At first, the manure itself causes animals to reject the herbage near it, probably because of unpleasant odor, but later the forage becomes too mature and then is unpalatable because of its coarseness. In poorly managed pastures, herbage around manure patches may be rejected for as long as 18 months. In well managed pastures, in which soil life has become enlivened as a result of grazing management, manure can be disintegrated and incorporated with the soil within about 60 days, and the zone of repugnance disappears.

During the grazing season each cow deposits about 38, 19, and 10 pounds of nitrogen, phosphate, and potash in manure somewhere in the pasture. During the same time, each horse drops 40, 22, and 29 pounds of the same nutrients in manure, usually in very limited areas of the pasture or paddock. These nutrients are worth a lot of money and can be extremely beneficial to the pasture and your bank account, but not when the manure is stacked up. It must be spread around. Besides releasing nutrients, spreading exposes more of the manure to sunlight, which kills parasites and reduces breeding sites for flies. How it's spread is up to you, the pasture manager.

Under Voisin controlled grazing management of cattle, the high stocking densities and short grazing periods of paddocks usually result in uniform grazing and distribution of manure, and breakdown of manure by hoof action of the densely stocked animals. In New Zealand, where grazing is planned and controlled at a level beyond anything done in the USA, mowing or harrowing routinely after cattle to clip uneaten forage or spread manure isn't done because it's unnecessary. Pastures are harrowed sometimes during the dry season, however, when cattle graze forage higher in dry matter that results in drier manure which doesn't beak down easily. On New Zealand horse farms, beef cattle at high stocking density follow horses through paddocks to graze uneaten forage and break down horse manure (Alan Henning, personal communication). So it is possible with management to eliminate or minimize the need, and therefore the expense, labor, and fuel consumption of mowing and harrowing to remove uneaten forage and spread manure after cattle.

On rough land, management is your only option for spreading manure and maintaining the sward in good condition. Part of that management can include such things as grazing cattle, horses, sheep, goats, pigs, chickens, or turkeys together or one group behind the other. It can also include encouraging birds to feed in your pasture by putting up birdhouses for them to nest in. Birds digging and scratching in manure for insects help break it down and spread it around.

Horses are a special problem because high stocking density doesn't work very well with them. Three possible ways of keeping horse pasture forage in good condition and breaking down and spreading horse manure are: 1) follow horses through paddocks with heavily stocked cattle or sheep, 2) mow paddocks with a rotary mower, and/or 3) drag paddocks with a flexible tine harrow.

If the pastureland is level to rolling, you can use machinery to mow uneaten forage and spread manure if necessary (e.g. your grazing management isn't what it could be, or you're grazing horses without following with cleanup grazers). Using a rotary mower breaks up and spreads manure somewhat. If you use a sickle mower or don't mow, it's an especially good idea to drag paddocks with a flexible tine harrow in spring, fall, and two or three times during the season to break up and spread manure.

Urine

The area of pasture affected by urine varies with soil moisture conditions, slope, and type. Urine spreads further in wetter soil, down slopes, and through sandy soils than it does in dry, level, or clay soils. Provided that affected plants are not allowed to grow rank, animals don't seem to avoid eating plants growing in urine patches. In fact some animals prefer plants growing in urine patches, if they weren't burned by the urine. If urine scorches or kills plants, animals avoid eating them, and the botanical composition declines in quality.

SEED DISPERSAL

Seeds can attach to hooves, hides, and wool, and be spread around a pasture as animals graze. Seeds can also be carried around in digestive tracts of livestock, and then dropped in manure. Passage of undigested seeds through ruminants takes from 12 hours to about 6 days, depending on the animal. The degree to which eaten seeds are digested depends on both animal and plant species. Some animal digestive systems break down seeds more than others. Some seeds have harder seed coats that resist digestion more than others. About 10 percent of the seed that's eaten passes unharmed through animals.

Large numbers of seeds can be spread around by animals in their manure (10 live white clover seeds per gram of manure dry matter have been found). If these are seeds of desirable plants it benefits the pasture, but if they are undesirable they will only contribute to making the pasture more weedy. For this reason, it is always best to feed only forage that is relatively free of weeds. Weed seeds in hay that is fed during winter may pass through unharmed in manure, which is then spread on the land, creating weed-control problems. If possible, don't allow your livestock to graze a weedy pasture when the weeds have set seed, before grazing a clean pasture. Of course, if you manage your pasture well, weeds will never flower and set seed!

SUCCESSION

Plant and animal (including soil organisms) succession moves naturally toward greater complexity and stability. In northeastern and northcentral USA and similar areas, energy (sunlight, temperature, precipitation) resources are high enough to support complex forest vegetation. Once forest is cleared, there's surplus energy available. If it's not all used for pasture plant production, the extra energy will be used to grow woody plants and weeds. Maintaining pastureland under these conditions requires periodic effort or work to hold back the successional movement to climax forest. The larger the energy resources of a region, the more frequently work must be done, or the more severe it has to be if done infrequently to bring the plant and animal community back to a young growth stage. You can accomplish this either through grazing management with frequent moves of livestock at high stocking density and high utilization of pasture, or infrequent use of bulldozers, chain saws, and herbicides. The former method puts money in your pocket; the latter takes money from you.

REFERENCES

Gooding, A. 1980. Don't let your pastures waste money. Angus Journal. June-July. p. 348-350.

Johnstone-Wallace, D.B. and K. Kennedy. 1944. Grazing management practices and their relationship to the behaviour and grazing habits of cattle. J. Agric. Sci. 34:190-197.

Harris, W. 1978. Defoliation as a determinant of the growth, persistence and composition of pasture. p. 67-85. In Wilson, J.R. (ed.) *Plant Relations in Pastures.* CSIRO, Melbourne, Australia.

Savory, A. 1988. *Holistic Resource Management.* Island Press, Washington, D.C. 564 p.

Semple, A.T. 1970. *Grassland Improvement.* Leonard Hill Books, London. 400 p.

Smetham, M.L. 1973. Grazing management. p. 179-228. In Langer, R.H.M. (ed.) *Pastures and Pasture Plants.* A.H. & A.W. Reed, Wellington, New Zealand.

Smith, B., P. Leung, and G. Love. 1986. *Intensive Grazing Management.* The Graziers Hui, Kamuela, Hawaii. 350 p.

Vallentine, J.F. 1990. *Grazing Management.* Academic Press, New York. 533 p.

Voisin, A. 1959. *Grass Productivity.* Philosophical Library Inc., New York. 353 p.

Watkins, B.R. and R.J. Clements. 1978. The effects of grazing animals on pastures. p. 273-289. *In* Wilson, J.R. (ed.) *Plant Relations in Pastures.* Commonwealth Scientific and Industrial Research Organization, East Melbourne, Australia.

Voisin Controlled Grazing Management

Sitting silently, doing nothing,
Spring comes,
And the grass grows by itself.

Zenrin

I don't think anyone would fill a bunk or trench silo with corn or alfalfa ensilage, and then turn livestock into it, allowing them to eat what and when they want to. Of course no one would do that intentionally, because everyone knows what would result: a mess! Instead the silage is carefully rationed out according to the needs of livestock, while keeping waste of the feed to a minimum.

But everyone who has grazed animals, including me, has turned them out into large pasture areas, allowing them to pick and choose, eating what and when they want to. The same kind of mess results. The animals tend to eat plants and parts of plants that they like best, leaving the rest to mature, set seed, and multiply. The most desirable plants, such as clovers, are grazed off every time they grow to grazing height. They never have enough time with enough leaf surface for photosynthesis to meet the needs of the plant. As a result, these plants soon wear out and die. With this kind of management, any pasture becomes weedy and unproductive. But in the United States it hasn't been recognized as a management problem; instead, the pastures have been blamed. Not many people would try to grow corn, soybeans, wheat, or alfalfa with the same level of attention to cropping management and soil fertility that they use on their pastures. If they did, yields of those crops wouldn't amount to much of anything. But permanent pastures in the United States traditionally have been managed according to the Back-40 Syndrome: don't put anything in, don't get anything out.

The Back-40 is the run-down permanent pasture where animals are placed in spring and survivors are collected in the fall. If most survive, the farmer congratulates him/herself for doing a good job of management. Some animals even gain weight if the pasture was well understocked. When animals requiring a high level of nutrition, such as milking cows or growing lambs, are grazed with similar management, they aren't very productive.

It's no wonder that farmers in the United States stopped pasturing their animals and went to year-round confinement feeding: the yield of livestock products from pasture was too low to be profitable. But the ironic part is that it never was the pasture's fault. It always resulted from poor management of a forage crop in a pasture situation.

If it weren't for New Zealand's highly productive and profitable agriculture, which depends almost entirely on permanent pastures, American farmers probably would have gone on forever, blissfully ignorant of better grazing management techniques and spending much more than they need to in producing livestock. But New Zealanders produce lamb and dairy products, ship them halfway around the world, and underprice American farmers. They are doing something right. But what?

New Zealanders adapted the ideas of Andre Voisin to their conditions with great success. Voisin was a biologist and chemist, who taught at the National Veterinary School of France and at the Institute of Tropical Veterinary Medicine in Paris. He also was a Laureate Member of France's Academy of Agriculture, and held an Honorary Doctorate degree from the University of Bonn, Germany (another French citizen honored in this way was Louis Pasteur). Voisin also farmed in Normandy, France and he remained essentially a farmer in his outlook. That is, he was able to observe and understand natural interrelationships that often are missed by people not as perceptive and in tune with nature as farmers are.

Combining his scientific training and experience with diligent observation of livestock in the field, Voisin discovered that time was crucial in grazing management. Like many people who have made important discoveries, Voisin didn't receive the credit and recognition that he deserved, especially in the United States. Without the concept of time we still wouldn't know how to properly manage grazing livestock.

When Voisin defined the concept of time in grazing management, some important things were realized, especially about overgrazing and stocking density. It became possible to minimize overgrazing and undergrazing. He showed that overgrazing is unrelated to the number of animals present in a pasture, but is highly related to the time period (how long and when) during which plants are exposed to the animals. If animals remain in any one area for too long and graze regrowth, or if they return to an area before previously grazed plants have recovered, they overgraze plants. For example, 200 cows grazing a 1-acre paddock (that has adequately recovered from a previous grazing) for 12 hours don't overgraze, but one cow grazing the same 1-acre paddock for 7 days does overgraze.

Voisin called his flexible rotational grazing management method "Rational Grazing", because with it pasture forage is rationed out according to the needs of the animals (just as feed is rationed out in confinement feeding), while protecting the plants from overgrazing and achieving a high level of forage utilization.

Voisin's method of planned and controlled grazing management interferes as little as possible with the pasture environment, while gently guiding it to benefit the farmer, and protecting it from damage by grazing animals. It is a simple method that in essence just gives pasture plants a chance to photosynthesize and replenish energy reserves after each grazing. Using this method, you control what and when livestock eat, by dividing pastures into small areas (paddocks) and rotating animals through them. You can then ration out pasture forage according to the needs of livestock, allow the plants to recover from grazing according to their needs, and keep forage waste to a minimum.

Key parts of Voisin controlled grazing management concern rest or recovery periods between grazings, and the length of time that animals are in a paddock. Let's look at those parts and related aspects.

TERMINOLOGY

First I want to introduce some useful terms that can help in discussing and understanding what's going on in a pasture.

Pasture mass: the total weight of forage per acre, measured to ground level, and expressed as pounds of dry matter per acre (lb DM/acre). When measured before grazing it's called "pregrazing pasture mass"; measured

100

after grazing it's called "postgrazing or residual pasture mass".

Forage allowance: the total dry weight of forage, measured to ground level, that's offered to animals during a grazing period; expressed per animal or per unit of animal liveweight (e.g. lb DM/cow or lb DM/100 lb liveweight).

Net forage: the amount of forage that actually can be harvested from a pasture (lb DM/acre), either as forage consumed by grazing animals or removed by cutting. Rate of net forage production represents how much of that harvestable amount is produced per day (lb DM/acre/day).

Stocking rate: the number of animals carried or supported per acre during a season or part of a season (e.g. 1.4 cows/acre). It is directly related to the amount of feed grown per acre. Stocking rate is extremely important in balancing the feed requirements of livestock with the amount of pasture forage grown on a farm (see Chapter 8).

Stocking density: the concentrated number of animals grazing a paddock at a given moment, expressed as number of animals per acre per time (e.g. 200 cows/acre/12 hours). Stocking density greatly influences pasture plant growth, forage utilization, and through its effect on the herd's or flock's level of feeding, also affects the animal's feed conversion efficiency.

PASTURE SWARD DYNAMICS

To manage pastures well, the biological and ecological bases for management must be understood. Then grazing management can be flexible, based on observation of plants, soil, and animals, and understanding what's going on, rather than rigidly following a set schedule of calendar dates that may have very little or nothing to do with pasture plant growth.

Basically, it all boils down to the fact that pasture plants must be able to regrow after they have been grazed. The regrowth is powered by energy either from photosynthesis occurring in remaining leaf surface, or from energy reserves if little or no leaf surface remains. Except for alfalfa (which regrows from energy reserves) and after drought, pasture grasses and legumes can regrow mainly from photosynthesis occurring in remaining leaf surface. Regrowth from energy reserves is slower because it takes time to form enough leaf surface so that photosynthesis can function at a high rate again. Photosynthesis supplies energy for continued regrowth of the plant and storage of more energy reserves. Low stocking density and poor forage utilization (lax grazing) result in sparse swards having few or no lower leaves on plants because of shading; when grazed closely, such plants regrow slowly because little or no leaf surface remains. So if plants are cut or grazed before enough reserves are stored, and there are few or no lower leaves remaining, regrowth will be retarded or won't occur at all.

Plant Tissue Flow

Underlying that simple overview above is a complex interrelationship of plants, temperature, light, soil, organisms, nutrients, water, and livestock that makes pasture a continually changing (dynamic) community. In a pasture there is a continuous flow of new plant tissue forming and old tissue disappearing through the processes of aging, death and decay, or consumption by grazing animals (Figure 5-1). This turnover of tissue can occur very fast. For example, in perennial ryegrass a new leaf appears on each tiller about every 11 days. Ryegrass plants maintain only three live leaves per tiller, so each leaf lives an average of just 33 days. If leaves aren't eaten within their short lifespan, they're lost to grazing animals.
Climate, soil fertility, and plant species mainly

determine the rate of new herbage formation in a pasture sward. Grazing management can influence this rate by maintaining as green and leafy a sward as possible, so photosynthesis can occur all the time, including right after grazing when the plant residue is very short. Sparse, stemmy, yellow postgrazing plant residue with little green leaf surface area remaining, takes longer to reach a maximum rate of new herbage formation than does a dense, leafy, green residual.

Plant Regrowth Curve

The regrowth curve (Figure 5-1) of plants is S-shaped and has three stages: 1) early period of slow growth, 2) middle period of rapid growth, and 3) final period of slow growth. In the first growth of spring or after being grazed or cut off anytime in the season, plants have limited leaf surface area and can only grow slowly. As leaf surface develops and light interception increases in full sunlight, pasture plant growth increases quickly, then gradually slows as pasture mass and shading increase. The rate of new plant growth reaches a maximum at a pasture mass of about 800 to 1100 lb DM/acre in dense, leafy swards grazed by sheep. Pasture grazed by cattle is less dense and more clumpy, with more erect tillers and leaves than sheep pasture. Because less light is intercepted, cattle pastures need more leaf surface area and a pasture mass of 2200 to 2700 lb DM/acre to reach maximum growth rate (Figure 5-1).

Net Forage Production

After grazing, the rate of net forage production increases rapidly as plant growth and pasture mass increase, but at high levels of pasture mass, net forage production decreases as pasture mass continues to increase. This decrease reflects the smaller proportion of new plant growth being harvested, as more forage is lost through death and decay from shading of lower plant parts and low-growing plants such as white clover. Shaded plant parts continue to respire, but without light they can't photosynthesize, so until they die they draw energy from parts exposed to the sun. This is a direct loss to your pasture's productivity, and is the reason why swards must be kept below 6 to 8 inches tall!

Sooner or later as pasture mass continues to accumulate under poor or very lax grazing management, net forage production becomes zero (i.e. all of the forage rots and animals won't eat it), and pasture mass can't increase any more because of shaded conditions within the sward. There is a range of pasture mass over which net forage production remains high. The rate of net forage production drops significantly only at the extreme high and low levels of pasture mass (Figure 5-1).

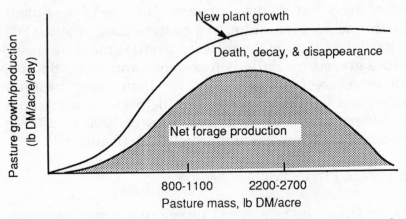

Figure 5-1. Influence of pasture mass on rates of new plant growth, net forage production, and forage loss through death, decay, and disappearance (Adapted from Korte, Chu, and Field 1987).

RECOVERY PERIODS

Besides variation of the plant regrowth curve, plant growth rate also differs within the season. One of the main rules of Voisin controlled grazing management is that recovery periods between grazings must vary according to changes in plant growth rates, which reflect changes in growing conditions. In general, this means that recovery periods must increase as plant growth rate slows as the season progresses. For example, in the northeastern and northcentral United States plant growth rate in May through June is about twice as fast as it is in August through September, and July is a transition time between the two. This means that recovery periods between grazings must be about twice as long in August-September as they are in May-June. Of course, plant growth rates vary within regions and prevailing climatic conditions in any season. (Figure 5-2)

Productivity of plants and the amount of forage available to animals entering a paddock, equals the daily amount of plant regrowth per acre (pasture mass) that accumulated since the last time the paddock was grazed. Looking closely at Figure 5-2, you can see that if the recovery period is cut to half of the optimum period, forage accumulation is reduced about two-thirds. If the recovery period is shortened even more, forage accumulation may drop to only 10 percent of the pasture mass accumulated during the optimum recovery period. This short recovery period corresponds to what happens when pastures are continuously grazed: desirable plants are grazed off every time they grow tall enough to be grasped by the animals' mouths, or about every 7 days. If recovery periods are longer than the optimum, pasture mass increases, but the increase is due mainly to more fiber, which lowers the feeding value of the forage.

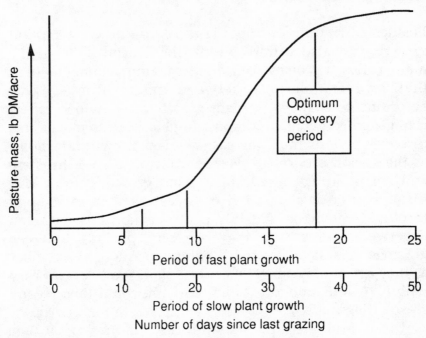

Figure 5-2. Relationship of recovery period to pasture mass accumulation during periods of rapid (e.g. May-June) and slow (e.g. August-September) plant growth (Adapted from Voisin 1959)

Basic recovery period guidelines are helpful for planning and beginning to use controlled grazing management. As you gain experience, you can make adjustments to better suit your local conditions and pasture plant communities. For example, in the Champlain Valley of Vermont, we have found that these recovery periods work well for grass-white clover pasture: 12 to 15 days in late April to early May, 18 days by May 31, 24 days by July 1, 30 days by August 1, 36 days by September 1, and 42 days by October 1.

Recovery periods needed in other climatic zones or for other pasture swards (e.g. grass-alfalfa) can be determined by taking into account what is known about the plants' carbohydrate reserve cycles, experimenting with different periods, and observing the effects of different pre- and postgrazing pasture masses on plant regrowth rates.

Recovery periods are, of course, based on observation of plant regrowth and pasture mass. Under conditions that stress plants, such as drought or cold, longer recovery periods are needed. If conditions are more favorable (e.g. warm, moist) than usual for plant growth, less recovery time may be needed. For example, depending on the animals and their production levels and nutritional needs, we allow paddocks to accumulate pregrazing pasture masses of 2200 (for sheep) to 2700 (for cattle) lb DM/acre in each rotation, before animals are turned in. We remove animals from paddocks when the swards are grazed down to 900 to 1350 lb DM/acre.

The days of recovery are a guideline to facilitate planning and management. Watch the plant regrowth and pasture mass: they are your best indicators of what conditions are like for the plants. Movement of animals among paddocks must always be based on observation of plant regrowth and pasture mass. By observing this need for adequate recovery time between grazings, pastures in areas such as Vermont's Champlain Valley can be grazed from April 15 to November 15, or longer!

If the length of time needed to graze a paddock to the required residual pasture mass decreases, the recovery periods for the next paddocks also will be shortened if animals are moved sooner than planned. If recovery periods begin to shorten by even 12 hours in any rotation, it as a warning signal that plant growth has slowed for some reason. For example, around June 20 you notice that recovery periods, which had been 22 days, begin to shorten by 12 hours in each paddock. Carefully check the pasture mass in paddocks that the animals will go into next. If the plants haven't regrown enough in the next paddock, you must slow the movement of animals through the rotation to allow plants more recovery time. You can accomplish this by 1) increasing the pasture area and number of paddocks available for grazing, 2) removing all animals from the pasture and feeding them elsewhere, or 3) feeding the animals hay or other forage in a paddock or paddocks until enough pasture mass accumulates (2200 to 2700 lb DM/acre) and recovery periods are adequate again.

(Anytime that you feed hay, greenchop, or silage to animals, including during winter, do it on different areas of your pasture. If possible, feed on a different area every day. This improves the pasture soil fertility from the added manure, urine, and wasted forage, saves you the time and expense of cleaning up and spreading manure, and adds new forage seeds to the pasture.)

Another way to increase recovery periods early in the season is to graze areas that had been set aside for machine harvesting, even though the surplus forage hasn't been cut yet. The animals won't like having to graze tall, more mature plants, after being accustomed to grazing short immature plants, and they will waste a lot of forage if you let them. To prevent waste, give them narrow strips of forage about 10 feet wide, or only enough for 12 to 24 hours. Hold the animals in the strips until they eat most of the forage. Then if possible mow each strip so that they can eat the remaining forage before giving them another strip. It's hard for ruminants to bite off stemmy plant parts, so you help them by mowing the plants in this situation.

Animals needing high levels of nutrition (e.g. milking cows, sheep, or goats, growing lambs) can't be forced to graze this more mature forage completely without reducing their production, however. So don't force animals with high nutritional needs to graze it as closely. But mow the area that is used for grazing while increasing recovery periods, at least once, either in every strip or as soon as you remove the livestock from the area. Mowing cuts the stems that weren't eaten and evens the plant growth for grazing in the next rotation.

If you have to increase recovery periods later in a season, your animals can graze hayland aftermath, sudangrass, millet, or other crops seeded for this purpose. You also can feed green chop or feed some of the surplus forage harvested earlier from the pasture. Remember to feed machine-harvested forage in different areas of a paddock or paddocks, if possible.

If plants in the next paddocks to be grazed don't recover enough, less forage is available and, consequently, movement of animals accelerates through the rotation at a time when it should be slowing down. Voisin referred to this faster movement of animals as "untoward acceleration". Suddenly the plants become exhausted, stop growing, and there's no more forage to graze.

An important question that must be answered is: how can the length of recovery periods be adjusted during the season until they are about twice as long in autumn as in spring? There are three practical ways for doing this:

1. Usually about half of the pasture area must be set aside from grazing in the spring (or other periods of rapid growth) and machine-harvested, because too much forage is produced in May-June. This means that only about half of the total pasture area is grazed in May, June, and part of July! Because using controlled grazing management at least doubles or triples plant productivity, you will have to machine-harvest forage from areas where surplus forage probably has not been harvested before. If your pastures are mainly on rough land, set aside the most level areas where you can use machinery to harvest the excess forage. You must be prepared for the increase in forage production that will occur, otherwise the pasture won't be grazed properly and its full potential won't be realized. After surplus forage has been harvested, allow the areas to recover until an adequate pasture mass has accumulated,

divide them into paddocks, and include them in the rotation. This increases the area available for grazing and automatically lengthens recovery periods to about what is needed for the remainder of the season.

2. Graze twice as many animals on the total pasture area in May, June, and the first part of July as in the rest of the season. The decrease in animal numbers carried during the second half of the season lengthens recovery periods to about what is needed. This really is the only way to graze a pasture well if it is all rough land that can't be harvested with machinery anywhere. This means that about half of your animals will either have to be fed elsewhere after July 15, or you will have to sell them.

3. A compromise way of keeping pasture forage under control during times of excessive growth, on land where machine harvesting isn't possible, is to not graze down as low as usual, leaving more postgrazing pasture mass. This means that animals will rotate through the paddocks more quickly than usual. The patchy grazing that probably will occur can be minimized by grazing two kinds of animals (e.g. sheep and cattle) on the same land, either simultaneously or one behind the other.

PERIODS OF STAY AND OCCUPATION

The length of time that each group of animals is in a paddock per rotation is called the period of stay. The total time that all groups of animals occupy a paddock in any one rotation is called the period of occupation. If only one group of animals grazes a paddock system, the period of stay equals the period of occupation. If two groups graze, their total periods of stay equal the occupation period.

When plants are grazed off, they can regrow tall enough to be grazed again in the same rotation if the occupation period is too long. So periods of stay or occupation must be relatively short to prevent grazing of regrowth. In the

110

northeastern and north central United States, plant regrowth may be tall enough to be grazed again after about 6 days in May-June and 12 days in August-September. Although most plants take 12 days in August-September to regrow to grazing height, some continue to regrow tall enough to be grazed after only 6 days. For that reason, periods of occupation must never be longer than 6 days to prevent grazing of regrowth in the same rotation, and really should be 2 days or less for best results.

Periods of stay for any one group of animals shouldn't be longer than 3 days, giving total occupation periods of 6 days for two groups of animals. This is because the longer animals are in a paddock, the less palatable the remaining forage becomes, and the more time and energy they spend searching for desirable feed. Periods of stay of 2 days or less if animals are grazed as one group, and 1 day or less for each of two groups, giving total occupation periods of 2 days or less, are better than longer periods of stay and occupation.

In practice then, the shorter the periods of stay and occupation, the better the conditions are for optimum plant and animal production. Milking, growing, and fattening animals shouldn't be in a paddock for longer than 2 days per rotation anytime in the season, to keep them on a consistently high level of nutrition. Milking cows, sheep, and goats produce the most milk if they are given a fresh paddock after every milking. Not only is the forage of higher quality and grazed more uniformly than with less frequent moves to fresh paddocks, but milking animals let their milk down better, anticipating that they're going to a fresh paddock as soon as they're done milking. Growing and fattening animals, such as lambs and beef animals, also gain weight most rapidly if given a fresh paddock every 12 or 24 hours.

Paddocks must be small enough so that all forage in each paddock is grazed completely and uniformly within each occupation period. Occupation periods may need to change because plant growing conditions vary during the season, and the amount of forage available changes. Lengthening occupation periods and pregrazing pasture masses consistently above 2700 to 3100 lb DM/acre indicate that surplus forage is available. When that happens, some paddocks should be removed from the rotation and machine-harvested, so that the pasture continues to be well grazed to maintain forage quality, sward density, and plant regrowth potential. Shortening occupation periods and pregrazing pasture masses below about 2000 to 2200 lb DM/acre indicate that plant growth rate has slowed and not enough forage is accumulating: more paddocks and pasture area are needed.

For example, if animals don't eat enough to keep up with the rapidly growing plants in May and June, remove paddocks from the rotation and cut the forage for hay or silage. Suppose that in the first rotation of the season your animals occupy paddocks for 2 days, eating all of the forage available in each paddock within the 2 days. After they have grazed six or seven paddocks, move them to the first paddock that was grazed and start the second rotation. Leave the rest of the pasture area for machine harvesting. (Return animals to the first paddock only if the plants in it have regrown to at least 4 inches tall and have a fully developed green color. Never allow animals to graze forage that's so young that it is still yellow. Grazing young yellow plants has a similar effect on animals that eating green apples has on people! Grazing plants that are too immature also may result in the plants growing poorly during the rest of the season.)

Recovery periods and occupation periods should work together so that the forage is at the right pasture mass or height when animals are turned into a paddock. When animals enter a paddock, plants should be about 6 inches tall (about 2500 lb DM/acre pasture mass) for cattle or horses, and 3 to 4 inches tall (about 2200 lb DM/acre pasture mass) for sheep, goats, pigs, or poultry. These pregrazing plant height differences are necessary because plants are not as dense in pasture grazed by cattle as they are in pasture grazed by sheep or goats. They also reflect the need to provide forage with physical characteristics that facilitate grazing and intake. Cattle can graze 6-inch tall forage efficiently, but sheep and goats require shorter forage to graze it well.

When animals leave a paddock, ideally all or most plants should have been grazed down to about 1 to 2 inches from the soil surface (900 to 1300 lb DM/acre pasture mass). Notice the word "ideally" in the last sentence. We are working with animals that don't seem to have the slightest idea that we would like them to graze closely and uniformly. Their natural inclination is to eat the most palatable forage and leave the rest, especially forage anywhere near some of their own manure. Depending on plant species and local environment, postgrazing plant heights and pasture masses may need to be greater. But usually the taller the remaining plants are when animals are removed from a paddock, the more selective and uneven the grazing has been. In the early stages of pasture improvement, it may be very difficult to have animals graze closer than 2 inches from the soil surface, because of stubble and matted plant parts from previous years of poor forage utilization. It always will be difficult to get animals to graze closely in areas that have been machine-harvested, probably because the dry fibrous stubble remaining after mowing picks the animals' mouths. One way to avoid leaving stubble is to machine harvest surplus forage twice, or every time the sward reaches about 10 inches tall, rather than once when the sward is 18 or more inches tall. Two or more immature harvests result in leafier hay or silage, and less loss of low-growing plants from shading. Of course, if it's difficult to

113

dry early cut forage in your area, you may be limited to one harvest taken later. One way to delay excess forage harvests until there are better drying conditions, is to graze surplus areas in the first rotation, then set them aside for later cutting(s).

Animals with high nutritional needs (e.g. milking cows, sheep, or goats) probably shouldn't graze forage down to 1 inch (900 to 1100 lb DM/acre postgrazing pasture mass) from the soil surface, unless the sward is very dense and leafy because their production may be lowered. This is why it is best to follow producing animals with animals having lower nutritional needs (e.g. dry cows, heifers, dry ewes) to graze paddocks down closely (discussed below). If for some reason you can't graze with two groups of animals, don't force producing animals to graze down to 1 inch unless you are willing to lose some production. If your pasture can be mowed, producing animals can be allowed to graze less closely, and the pasture can be mowed after each grazing or periodically during the season. This is a compromise though, because, while your animal production will remain high, the pasture won't reach its full potential that ultimately could have resulted in even higher levels of animal production. Running machinery to mow or do anything else also takes time, costs money, and burns fuel.

Just as you must protect the pasture from being overgrazed by using short occupation periods, you should make certain that the pasture isn't undergrazed as well. Any plants that are not grazed in one rotation, probably won't be grazed again that season unless they are clipped. It saves you time and money, and net forage production increases if the animals graze as uniformly as possible down to 1 to 2 inches from the soil surface.

STOCKING RATE AND DENSITY

High stocking density, coupled with quick, close grazing of a paddock, and removal of animals until plants recover imitates what happened under natural conditions. Then large herds of grazing animals grouped closely because of predators would move through an area eating and trampling plants and disturbing the soil. Their passage broke up plant residues and returned them to the soil to be decomposed. Disturbing the soil allowed seedlings to establish and water to enter. Large amounts of manure left behind provided nutrients for plant regrowth. By the time the herd returned the plants had recovered and were ready to be grazed again. Pasture and range plants evolved under this kind of treatment and require it for best growth. This is one of the underlying reasons why Voisin controlled grazing management has such beneficial effects on pasture.

It follows that stocking rate and density greatly affect pasture forage utilization and net forage production. Increases in stocking rate and density always reduce the amount of pasture forage that is wasted, thereby increasing the efficiency of forage utilization and net forage production.

On the most intensive and efficient dairy farms, for example, cows eat 80 to 90 percent of forage available to them during the season. On less well managed farms, half of the forage produced may not be eaten, mainly because of low stocking rate and density. Wasting pasture forage prevents getting high levels of plant and animal production per acre.

Increasing stocking rate and density results in pastures being grazed more intensively during most of the season. Benefits of intensive grazing include less dead and dying herbage in the pasture (i.e. net forage production goes up), better forage digestibility, and more white clover and tillering of grasses in the sward.

Of course, when you increase the stocking rate on your farm, more livestock graze per acre and per ton of pasture forage produced. Less pasture will be wasted, but eventually you could increase to a stocking rate where the amount of forage eaten per animal decreases, and its production level consequently declines. It's relatively simple to estimate what the stocking rate should be, based on animal energy and dry matter intake needs and pasture forage production (see Chapter 8). Pasture stocking rates generally range from 0.7 to 1.8 animal units (1 animal unit = 1000 pounds liveweight) per acre.

Stocking density depends on how much forage is available to animals in a paddock, how much of it you want them to eat, and how long you want them in the paddock. Paddock size and stocking density should combine so that animals don't have to be moved more frequently than twice a day, or less frequently than every 6 days, as discussed above. Keep in mind that more intensive grazing management produces the highest pasture and animal yields. Stocking densities generally range from about 25 to 200 animal units per acre per occupation period of a paddock.

FORAGE ALLOWANCE

Livestock must be offered amounts of pasture forage that are about two to four times more than what they will eat, depending on the animals production level and nutritional needs. This is to ensure that they are able to eat as much as they possibly can. For example, if a milking cow needs to eat 35 lb DM/day, she must be offered a forage allowance of 70 to 140 lb DM/day. She will leave 40 to 60 percent of the amount offered uneaten as post-grazing pasture mass. So livestock that need to eat large quantities of forage usually must be allowed to leave relatively large amounts of postgrazing pasture mass (1150 to 1350 lb DM/acre). Animals with low nutritional and dry matter intake requirements can be given a lower forage allowance and eat more of it by grazing to a lower postgrazing pasture mass (800 to 900 lb DM/acre).

SWARD MEASUREMENTS

Besides understanding the biological and ecological reasons for what you do, you must be able to estimate the amount of forage present in a sward before and after grazing. It's helpful to discuss sward measurements before getting into other details of controlled grazing management.

Plant Height, Density, and Pasture Mass

Plant height and density combined are included in pasture mass. For a particular pasture botanical composi-tion and time during the grazing season, pasture mass is useful for predicting plant and livestock performance.

Pasture mass interrelates with forage quality and palatability in affecting productivity of grazing livestock. High pregrazing pasture mass over 2700 to 3100 lb DM/acre occurs with low stocking density or infrequent grazing. These high levels of pasture mass can quickly result in decreased forage quality, patches of rank, low-quality uneaten forage, a layer of dead and decaying plant material at the base of the sward, more upright growing plants, patches with no live tillers, death of low-growing legumes, and a shift to undesirable plant species.

For maximum plant and animal production, pasture mass generally should fluctuate between about 1100 (postgrazing) and 2900 (pregrazing) lb DM/acre, depending on the kind of livestock. Occasional close grazings down to 800 to 1000 lb DM/acre are needed to maintain high forage quality and desirable sward composition.

The main objective of Voisin controlled grazing management is to keep pasture plants within the steep part of their growth curve, so that the rate of new plant growth always remains high (Figure 5-1). If they're kept within the steep part of the curve, regrowth occurs rapidly after plants are grazed. If postgrazing pasture mass reaches less than 800 lb DM/acre, plants have little leaf surface and are then in the low part of the curve, and regrowth occurs slowly until adequate leaf surface develops. For this reason, pre- and postgrazing pasture masses of each paddock must be estimated, in deciding when to move animals in and out of paddocks (discussed below).

Pasture mass usually is estimated visually (eye-balling) or by measuring plant height (sward surface height). Sward height measurements or estimates must be calibrated against actual measurements of the forage present in an area (discussed below). Sward height affects plant growth and death through shading and, consequently, influences net forage production. Sward height also affects forage intake and performance of grazing livestock. For example, maximum utilization of perennial ryegrass-white clover by grazing livestock occurs when swards are maintained at 1.5 to 2.5 inches tall for sheep, and 3 to 4 inches tall for cattle! For other cool-season grasses (e.g. Kentucky bluegrass, orchardgrass) combined with white clover, best pregrazing sward surface heights appear to be 3 to 4 inches tall for sheep, and 4 to 6 inches tall for cattle. Optimum postgrazing sward heights are 1 inch tall for low-growing grasses (e.g. perennial ryegrass, Kentucky bluegrass), and 2 inches for tall-growing grasses (e.g. orchardgrass, timothy, bromegrass).

Using sward height to guide grazing management actually isn't completely satisfactory, however, because it doesn't take into account variations in plant species and density. For uniform swards, such as nitrogen-fertilized perennial ryegrass growing under uniform conditions, sward height correlates closely with plant density and pasture mass. But sward height measurements may be misleading for estimating the pasture mass of complex swards that contain several grass species, depend on a legume for nitrogen, and grow under variable conditions, such as in most of the northern United States.

Since plant density is harder to estimate than plant height, visually estimating pasture mass requires experience and also some actual measurements of the forage to calibrate visual estimates. Visual density and height estimates may be checked against the actual total amount of forage present, by first estimating the pasture mass in pounds of dry matter per acre and measuring plant height within several 1- x 3-foot areas. Then cut the forage within the areas at the soil surface and place the forage from each area in a paper bag that you have noted your estimates on. Dry the samples in an oven at about 160 degrees Fahrenheit for 1 day, and weigh (in pounds) the dry forage. Convert the dry weights to pounds of dry matter per acre by multiplying each dry sample weight by 14520. Compare your estimates to the actual amount of forage present.

Easier and more accurate ways of estimating pasture mass, especially in complex swards, would be to take about 30 readings throughout a paddock with a pasture bulk height plate or with an electronic capacitance meter, both of which take plant density and height into account.

We have been using a pasture bulk height plate developed by Ed Rayburn (see references) that is very inexpensive and simple to make and use. Just buy an 18-inch square piece of 1/4-inch thick acrylic plastic and a yard stick at a hardware store. Cut a 1.5-inch diameter hole in the center of the plastic square. (Cutting the hole is the most difficult part, because the plastic tends to melt back together. Try to have someone cut the hole at the hardware store.) Attach a 1/8- x 2.5-inch bolt through a hole in the lower end of the yardstick, to pick up and carry the plastic plate easily as you walk around your paddocks. Now insert the yardstick through the hole in the plastic plate, and you're all set!

Start taking readings near one corner of a paddock in an area that will be grazed. Always estimate pasture mass in representative areas that animals will graze, not in rejected forage areas around manure. Gently lower the flat side of the plate with the upright yardstick onto the sward. Push the yardstick down until it just touches the soil, and allow the plate to settle by its own weight. Then bend over and look across the top surface of the plate to read the ruler measurement of forage bulk height. Record the reading to the nearest 1/4 inch on a paper or on a calculator that sums and tells you how many entries you've made. When you've finished taking about 30 readings by zig-zagging across the paddock, calculate the average forage bulk height. If your sward is composed of white clover and grasses such as Kentucky bluegrass, orchardgrass, timothy, bromegrass, and quackgrass, you can multiply the average bulk height by 432 to estimate pounds of dry matter per acre.

For best results the plate should be calibrated so that you know how many pounds of forage dry matter per acre the bulk height readings correspond to for your swards. To do this, make a wire frame that just fits over the plate. At several sample locations in your paddocks place the wire frame over the plate after you have taken and recorded a bulk height reading, and remove the plate. Then arrange the plants so that the wire frame lays flat on the soil surface. Plants rooted within the frame should be in the frame, and plants rooted outside should be outside of the frame. Then cut all of the plants within the frame at ground level with a scissors or battery powered lawn edger. Place the plant material in a paper bag, and note paddock number, date, and bulk height reading on the bag. Dry the samples in an oven at about 160 degrees Fahrenheit for at least 24 hours. Then weigh the samples and multiply the dry weights by 19360 if you weighed in pounds, or 42.6 if you weighed in grams, to get pounds of dry matter per acre.

Now graph the values, with inches of bulk height on the X-axis and pounds of dry matter per acre on the Y-axis. Future bulk height readings can be compared to the graph, to estimate dry matter per acre.

Another way of estimating pasture mass is with a computerized hand-held instrument that uses changes in electrical capacitance resulting from different plant surface areas, to measure pasture mass (e.g. Gallagher Pasture Probe). With the Pasture Probe, it can accurately be determined when paddocks are ready to be grazed, when animals should be removed from paddocks, and daily and seasonal forage yield of each paddock and the entire pasture. For best results, the Pasture Probe should be calibrated for your pasture sward, by taking Probe readings in measured areas, and cutting, drying, and weighing the forage from the areas.

(Unfortunately, the Pasture Probe has a few problems that can make using it frustrating. It lacks a battery charge indicator, so it's difficult to know when the battery needs charging before it shuts off while in use because of low charge. Some units apparently have a heat-sensitive component that causes them to shut off when they get too hot; because of lack of service backup, this problem has not been resolved. We have to keep the Probe computer in a cooler to use it! But despite those problems, it's a very helpful tool, and provides estimates of pasture mass with the least variation of all methods available.)

DIVIDING THE ANIMALS

The simplest way to graze is to include all animals in one group. Having the animals all in one group also gives the highest stocking density. Although only one source of water is needed when animals graze as one group, ideally drinking water should be provided in all paddocks so that animals, and their manure, remain in the paddocks being grazed. Also, if drinking water is always readily available in paddocks, animals don't waste energy walking to and from the source of water.

A more efficient way is to divide animals into groups according to their production levels and nutritional needs at different physiological states. This allows you to closely match pasture feeding value with animal needs. A dairy herd can be divided into two groups of 1) milking cows and 2) dry cows and heifers, or 1) milking does and 2) dry does and kids. A sheep flock can be divided into two groups of 1) weaned lambs and 2) dry ewes. Other kinds of livestock also can be divided into groups to improve efficiency.

The groups of animals can be handled in two ways:

1. Each group can be grazed in separate cells. (A cell is an area of land planned for grazing management as one unit, to be subdivided into paddocks to ensure adequate timing of grazing and recovery periods.) Animals with the highest nutritional requirements (e.g. milking cows, sheep, and goats, growing lambs or beef cattle) are grazed on the best pasture available. Other animals (dry cows and heifers, dry does and kids, ewes) are grazed in a separate cell on lower-quality pasture that's being improved. An advantage of this method is that only one source of water is needed in each cell, although providing water in all paddocks is preferable. Another advantage is that parasites aren't easily passed from older animals to young ones (this is especially important with pigs, sheep, and goats).

For producing or growing animals, paddocks should be small enough so that all forage is eaten in each paddock in 2 days or less (12 to 24 hours is best). Occupation periods for the animals at lower levels of nutrition can be 6 days or less, but they and the pastures will do better if 2-day (or less) periods of occupation are used also.

2. Both groups graze within the same cell. Animals having the highest nutritional needs are turned into a paddock first so they can eat the best forage quickly. They should not be left in a paddock for more than 2 days, because after that time they have to work too hard to meet their nutritional needs (12- to 24-hour periods of stay are best). After the first group is removed from a paddock, the second group follows to clean up the remaining forage, which has a lower feeding value than the forage that was grazed first. Paddocks must be small enough so that the combined periods of stay of the two groups are less than 6 days (2 days or less is best). When two groups of animals graze within the same cell, all paddocks must have a source of drinking water, because at least one of the groups must stay in its paddock to keep the groups separate.

NUMBER OF PADDOCKS

When planning for managing grazing, you need to estimate how many paddocks will be required to ensure adequate recovery periods between grazings. The number of paddocks needed depends on recovery periods, stay or occupation periods of animals in each paddock in each rotation, and number of animal groups grazing (Table 5-1). Since shorter occupation periods favor higher plant and animal yields, the more paddocks you have up to a certain point, the more productive your pasture will tend to be. In deciding how many paddocks to build, you should consider topography and soil fertility of the pastureland, pasture sward botanical composition and its potential yielding ability, maximum recovery periods likely to be needed in your area, your livestock, fencing costs, and your financial situation.

Table 5-1. An example of the number of paddocks needed for a 36-day recovery period between grazings.

Period of stay for 1 group, days	Total number of paddocks needed for	
	1 group	2 groups
1/2	73	74
1	37	38
2	19	20
3	13	14

First, estimate the recovery period likely to be needed during the time of slowest plant growth in your area's grazing season. This may range from less than 36 to more than 100 days, depending on conditions. For example, while 36 to 42 days recovery time are adequate for late summer and fall grazing in the northeastern United States, a 90-day recovery period may be needed in drier areas of other regions. Remember, these are the total numbers of paddocks needed when pasture plants grow the slowest. During times of fastest growth, such as in May-June in the Northeast, you will only need about half as many paddocks, since the rest of the pasture area will be set aside for machine harvesting.

Second, estimate how long the periods of occupation will be, and decide whether to use one or two groups of animals.

Third, use this equation to calculate the number of paddocks needed: recovery period/occupation period + number of animal groups = number of paddocks needed. For example, if 36-day recovery periods will be required, and you graze one group of animals with 12-hour occupation periods, the number of paddocks needed will be: 36/0.5 + 1 = 73. With one group and 1-day occupation periods, the number of paddocks needed will be: 36/1 + 1 = 37. If one group grazes with 2-day occupation periods, then: 36/2 + 1 = 19 paddocks needed. If you divide your animals into two groups and graze within the same cell, just add one more paddock: 36/0.5 + 2 = 74; 36/1 + 2 = 38; or 36/2 + 2 = 20.

The more paddocks that can be formed to provide adequate recovery periods and short occupation periods, the better. Up to about 30 paddocks, each new division significantly shortens the average occupation period of all paddocks. Beyond that, the gains become less as more paddocks are formed. Don't even attempt to control grazing with fewer than 10 paddocks, because the results will be very disappointing, and you may conclude that the method doesn't work. The more paddocks there are, the more flexibility exists for dealing with changing amounts of forage available during the season. For example, if you only have six paddocks, and for some reason the forage production in one of them drops off, the decline would affect 1/6 of your pasture area. In contrast, if you have 37 paddocks, a problem with one of them only affects 1/37 of your pasture area.

At this point, you may be thinking that building a lot of paddocks will be too expensive, and that the management required to control grazing will take too much time and effort. By trying to answer likely questions that may come up and to provide you with enough information to get started doing it, I may be making it seem more complicated on paper than it actually is in practice. Once you're managing grazing correctly, you'll see that any costs that were involved in getting started will soon be paid back from the increased profitability of your farm. You'll realize that the method requires much less time, labor, and expense than feeding animals in confinement. Because of recent developments in fencing materials, the cost and building of paddocks will be surprisingly low and easy to do.

The most important point of all this is that you have enough paddocks to provide the required adequate recovery periods, with the periods of stay and occupation that you plan to use. The main idea is to graze the sward at the optimum stage of growth for the plants and animals. Your plan should be thought of as a starting point that you build on and change as you gain experience from observing results, and as the pasture improves. Don't try to force the plants and animals to fit a rigid schedule. Remember that these are guidelines to help you develop a management routine of planning, monitoring, controlling, and replanning if necessary.

PADDOCK SIZE

After you decide how many paddocks to have, divide the total pasture area by the number of paddocks to roughly estimate the average area of each paddock. Paddocks don't have to be equal in area, but should provide more or less similar amounts of forage to facilitate moving animals on a fairly regular schedule, for your convenience. For example, pasture areas with fertile soils and good moisture conditions may produce twice as much forage as areas with poor, dry soils. Paddocks in the highly productive areas could be about half the size of those in the poor areas.

With electric fencing it's always possible and easy to relocate some fences or subdivide more if necessary. For example, as your pasture becomes more productive and the animals can't eat everything within the occupation period that you want to use (e.g. 12 hours), just divide the paddocks in half, in thirds, or whatever it takes to reduce the forage allowance enough. As stocking density goes up, there's more competition among the animals for feed and less selective grazing. Under heavy stocking density (e.g. 150 to 200 animal units/acre/12 hours), even high-producing dairy cows will graze more uniformly and close to the ground, and do well if the forage is high quality, the forage allowance is adequate, and their rations are correctly balanced for the desired level of production.

Paddock sizes must be adjusted according to the intensity of management that's wanted. I can't tell you an exact paddock area to use, because that depends on how often you want to move animals, on pasture productivity, and on numbers, kinds, and sizes of the animals grazing.

You can calculate what paddock areas should be from livestock dry matter intake needs and estimated pasture mass relationships (see feed planning in Chapter 8):

$$\text{paddock area} = \frac{\text{(no. of animals x dry matter intake)(days)}}{\text{pregrazing mass - postgrazing mass}}$$

For example, a herd of 70 dairy cows requiring 35 lb DM/head/day from pasture, are given a fresh paddock every 12 hours. If pregrazing pasture mass will be about 2500 lb DM/acre, and postgrazing pasture mass will be about 1300 lb DM/acre, then paddock area should be:

$$\frac{\text{(70 cows x 35 lb DM/head/day)(.5 day)}}{\text{2500 lb DM/acre - 1300 lb DM/acre}} = 1 \text{ acre}$$

Because conditions and situations vary so much, unless you do the above calculations, you'll just have to experiment with different sizes of paddocks. Start with 1/2- to 2-acre paddocks for cattle or horses, and 1/8- to 1/4-acre paddocks for sheep, goats, or pigs. The flexibility of portable fencing simplifies changing paddock sizes, and enables easy machine harvest of surplus forage from small paddocks, since subdivision fences can be removed.

TURNING OUT IN SPRING & ROTATION SEQUENCE

Two problems must be dealt with in the first rotation of spring: 1) the transition time needed by animals to adapt to eating green forage after months of eating relatively dry feed, and 2) the need to graze some paddocks a little too early so that others are not grazed way too late.

Bloat

Bloat can occur in pastured animals (mainly cattle) when they graze succulent legumes (alfalfa, red and white clover), especially in spring. Due to a complex interaction of animal, plant, and microbiological factors, a stable foam forms in the rumen. Because of the foam, gas can't escape. Pressure builds up in the rumen, causing it to swell and press against the lungs, preventing breathing. If not treated quickly, the animal dies.

Precautions are needed to avoid bloating and death of your animals, and to allow time for microorganisms in animal digestive tracts to adapt to using lush pasture forage that contains about 80 percent water:

1. During the first 2 weeks of spring turnout:

a) watch animals closely; call a veterinarian at the first signs of bloating;

b) fill animals up with hay before putting them on pasture for a couple of hours late in the morning or early afternoon (first few days);

c) move to fresh paddocks only when dew or rain has dried off;

d) provide dry hay to animals on pasture, in case some animals would like to eat additional dry fiber.

2. If for some reason you decide to apply nitrogen fertilizer to your pasture, don't do it in the spring because it results in forage being more succulent.

3. Feed antifoaming agents (e.g. Poloxalene, Bloat Guard) or add them to drinking water, beginning several days before the animals are first turned out to pasture, and continuing until they have adjusted to the lush forage.

4. Manage grazing so that the sward contains less than 50 percent clover or alfalfa. This can be achieved by allowing greater pregrazing pasture mass so that grasses shade the legumes and decrease their amount in the sward.

5. Tendency to bloat seems to be hereditary, so the best long term solution is to cull animals that bloat.

First Rotation Management

You'll have to begin grazing some paddocks when pasture mass reaches 1500 to 1800 lb DM/acre (plants are only 2 to 3 inches tall). But make certain that the plants have a fully developed green color in the first paddocks before beginning to graze. Unless you start early in the first paddocks to be grazed, by the time the animals reach the last ones to be grazed in the first rotation (12 to 20 days later, depending on occupation periods), the forage in them will be too mature, and the animals may not graze them well. This isn't as much of a problem if you are willing and able to clip the last few paddocks soon after they are grazed. Otherwise, the forage that's left in the first rotation will be there to haunt you all season, and will decrease the productivity of those paddocks.

Rotational Sequence

If you graze paddocks in the same sequence every year, you'll soon have as many different pasture plant populations as you have paddocks. This is because some paddocks will always be grazed too early for certain plants, some will be grazed just right, and others will be grazed too late. The way to distribute the stresses of early and late grazing in the first rotation, is to start with a different paddock each year and use the other paddocks in a different sequence, if possible.

During wet spring and fall conditions you'll have to graze higher and drier areas to avoid problems with mud, so it may be impossible to make much of a change in the sequence that your paddocks are grazed. But even if you alternate the first grazing each year between only two paddocks, it is better than not changing the sequence at all. As always, do what you can, knowing what the ideal situation would be.

Later in the season, if you notice that the forage in a paddock is ready to be grazed ahead of its time in the sequence, go ahead and graze it. Grazing out of sequence won't damage the plants, because their faster growth indicates that conditions in the paddock are more favorable than in other areas of the pasture, and that they have had adequate time to recover under those conditions. Grazing management must be flexible, to deal with the ever-changing pasture environment.

REFERENCES

Appleton, M. 1986. Advances in sheep grazing systems. In J. Frame (ed.) *Grazing*. British Grassland Society, Berkshire, England. 250 p.

Bingham, S., and A. Savory. 1990. *Holistic Resource Management Workbook.* Island Press, Washington, D.C.

Bircham, J.S., and C.J. Korte. 1984. Principles of herbage production. In A.M. Fordyce (ed.) *Pasture: The Export Earner.* New Zealand Institute Agricultural Science, Wellington, New Zealand.

Blaser, R.E., R.C. Hammes, Jr., J.P. Fontenot, H.T. Bryant, C.E. Polan, D.D. Wolf, F.S. McClaugherty, R.G. Kline, and J.S. Moore. 1986. *Forage-Animal Management Systems.* Virginia Polytechnic Institute & State Univ., Blacksburg.

Fletcher, N.H. 1982. *Simplified Theoretical Analysis of the Pasture Meter Sensing Probe.* Australian Animal Research Laboratory Technical Paper. CSIRO, East Melbourne.

Frame, J. 1981. Herbage mass. In J. Hodgson, R.D. Baker, Alison Davies, A.S. Laidlaw, and J.D. Leaver (eds.) *Sward Measurement Handbook.* British Grassland Society, Berkshire, England.

Griggs, T.C., and W.C. Stringer. 1988. Prediction of alfalfa herbage mass using sward height, ground cover, and disk technique. *Agronomy Journal.* 80:204-8.

Holmes, C.W., G.F. Wilson, D.D.S. MacKenzie, D.S. Flux, I.M. Brookes, and A.W.F. Davey. 1984. *Milk Production From Pasture.* Butterworths. Wellington, New Zealand.

Korte, C.J., A.C.P. Chu, and T.R.O. Field. 1987. Pasture production. p 7-20. In A.M. Nicol (ed.) *Feeding Livestock on Pasture.* New Zealand Society of Animal Production, Hamilton.

Nolan, T., and J. Connolly. 1977. Mixed stocking by sheep and steers: A review. *Herbage Abstracts.* 47(11): 367-74.

Parsons, A.J., and I.R. Johnson. 1986. The physiology of grass growth under grazing. In J. Frame (ed.) *Grazing.* British Grassland Society, Animal and Grassland Research Institute, Berkshire, England. 250 p.

Rayburn, E. 1988. The Seneca Trail Pasture Plate for Estimating Forage Yield. Seneca Trail Research and Development, Franklinville, New York. Mimeo. 5 p.

Reid, R.L., and G.A. Jung. 1973. Forage-animal stresses. p. 639-653. In. M.E. Heath, D.S. Metcalfe, and R.F. Barnes (eds.) *Forages.* Iowa State University Press, Ames.

Savory, A. 1988. *Holistic Resource Management.* Island Press, Washington, D.C. 564 p.

Sheath, G.W., R.J.M. Hay, and K.H. Giles. 1987. Managing pastures for grazing animals. p. 65-74. In A.M. Nicol (ed.) *Feeding Livestock on Pasture.* New Zealand Society of Animal Production, Hamilton.

Smetham, M.L. 1973. Grazing management. In R.H.M. Langer (ed.) *Pastures and Pasture Plants.* A.H. & A.W. Reed, Wellington, New Zealand.

Smith, B., P.S. Leung, and G. Love. 1986. *Intensive Grazing Management.* Graziers Hui, Kamuela, Hawaii.

Voisin, A. 1959. *Grass Productivity.* Philosophical Library, N.Y. Reprinted 1988. Island Press, Washington, D.C. 353 p.

Voisin, A. 1960. *Better Grassland Sward.* Crosby Lockwood & Son, London. 341 p.

Paddock Layout & Fencing

Over the river, the shining moon;
In the pine trees, sighing winds;
All night long so tranquil -- why?
And for whom?

Hsuanchueh

Right from the start, try to design an ideal paddock layout that requires minimal work to use and is as intensive as possible. You can use a less-intensive level of grazing management while the improved productivity pays for the rest of the setting up costs. You'll always be able to build toward the ideal layout, if you design it that way in the beginning.

You'll need to answer several questions: How intensive do you want your grazing management eventually to be? How much are you willing to spend in the first year to start the system? Which areas will be set aside in spring for machine harvesting? Which animals will need to return to a central handling facility (e.g. milking parlor), and how convenient must it be? Will you be able to provide drinking water to all of the paddocks?

How will you get animals through or around low wet areas?

Keep the design flexible for changes in production level and adding other kinds of livestock. In planning the layout, remember that fences cost less, are easier and faster to build, and last longer when built in straight lines.

TOPOGRAPHY AND HANDLING FACILITIES

The shape of paddocks and where lanes and dividing fences will be placed depends a lot on the topography of your pastureland, and its location relative to the barn or other handling facilities. This is especially important with dairy animals that have to be milked twice a day. Use an aerial photograph of your farm if possible, or make a sketch of it, to design the paddock layout. It's much easier to make changes if you draw paddocks and lanes on paper, before building any fences.

Try to form paddocks that are more square-shaped, rather than long and very narrow. This is because less fencing materials are needed to fence in a square-shaped paddock than a long narrow one, even though they contain the same area. (Here's a chance to use your high school geometry!) If your pastureland is level-to-rolling, though, you can divide it into large permanent rectangles with high tensile wire perimeter fences. Inside the rectangles use portable fencing, including fiberglass or plastic posts, rolling fence wheels (e.g. Gallagher tumblewheels), polywire, polytape, reels, and/or electric net fencing (see descriptions below under Fencing Materials and Construction). In this case, animals are given fresh areas of forage, ranging from square-shaped to long narrow rectangular or triangular strips, by moving portable fences within permanent rectangles.

Rough, Hilly Land

If your land is rough and hilly, as most permanent pastures tend to be, you should take slope aspect and location of hill crests into consideration when designing the paddock layout. For example, land with north-facing slopes should be in paddocks separate from land with south-facing slopes, if the areas involved are large enough to make a difference and to form separate paddocks. South-facing soils warm up quicker in spring, get hotter and dryer in summer, and stay warmer longer in fall than north-facing soils. So in early spring, south-facing slopes may be ready to graze 2 to 3 weeks before north-facing slopes. In midsummer, plants on south-facing slopes may grow very slowly because of the hot, dry conditions, while plants on north-facing slopes continue to grow rapidly. Plants on south-facing slopes can be grazed longer in fall than plants on north-facing slopes. Proper management of such areas is possible only if they are in separate paddocks.

Hill crests also should be contained in separate paddocks to promote uniform grazing and decrease nutrient transfer to favorite camp sites (see Chapter 4). Paddock divisions along hill crests force animals to graze sunny and shaded aspects of the hills, and prevent them from camping on sunny slopes after grazing the shaded slopes. (Figure 6-1)

Level-to-Rolling Land

If your pastureland is level-to-rolling, it's not necessary to permanently divide it into small paddocks. Instead, you can fence in large long, wide rectangular areas with permanent perimeter high tensile fencing, and subdivide inside with portable fencing that you move ahead of and behind the animals. This kind of pasture division saves a lot of money in fencing materials, and makes it easy to set

135

aside large areas of the pasture for machine harvesting of surplus forage. (see Figures 6-1, 6-2, 6-3, 6-4, 6-5)

For cattle and horses, build permanent one- or two-strand high tensile fences around rectangular areas up to about 600 feet wide and as long as you want. Subdivide the rectangles across their widths by using a polywire or polytape in front of the animals, supported either on light portable fiberglass or plastic posts or on tumblewheels. Rolling fence can be used to subdivide rectangles up to about 600 feet wide, but 300-foot widths are easier to work with. If portable posts are used, the rectangle shouldn't be more than about 150 feet wide for easy moving of front and back fences.

Follow animals with a polywire L-shaped back fence, also supported on portable posts, or on tumblewheels. The back fence prevents animals from eating regrowth, and the L-shape forms a lane for them to go to the barn or to water, if a water tank is not moved along with them. The lane formed by the back fence can be alternated each rotation between the two sides of a rectangle, so that one side doesn't get too worn and compacted or muddy. This capability of changing the lane location is especially important during wet spring and fall conditions.

Move the front polywire to give the animals only enough forage for the occupation period that you want to use. The advantage of using tumblewheels is that one person can very quickly give animals a triangular strip of forage by moving every other end of the polywire at each setting. Reels at each end of the tumblewheels make moves easier, because you can let out or wind up the polywire as needed. Another way is to use a reel with a friction brake at one end; this allows you to pull out the polywire as needed.

A front polywire supported on portable posts also works well, but requires a little more time and walking to move it. Portable posts that have a lip for stepping them in enable faster moving of the fences. A good way to move a front polywire on posts is to move one end first, and then reposition the posts to where you want them, without taking the polywire off the posts. Then just take up the slack with a reel at one end. Both tumblewheels and polywires on portable posts can be moved without turning the perimeter fence off. It's a good idea to disconnect the polywire/post front fence from the electricity, though, because it's tricky to move it without touching it. Clips are available for disconnecting polywires from perimeter fences.

The back fence doesn't have to be moved every time the front wire is moved, but the more frequently the back fence is moved the better. This is because the longer the animals have access to an area of pasture, the longer the occupation period is for that area. Occupation periods that are too long may allow grazing of regrowth in the same rotation, and the pasture will not reach its full potential. Short occupation periods also reduce treading damage when soils are wet. Many dairy farmers in Holland have found that moving front and back fences every 12 hours results in highest plant and animal productivity. Whatever you do, move the back fence frequently enough so that regrowth can't be grazed in the same rotation.

For sheep and goats, build rectangles that are multiples of 145 feet wide (e.g 145, 290, 435, 580, etc.) by as long as you want. The 145-foot increments are needed because the electric net fencing for small livestock comes in 150-foot rolls. So why not make the rectangles in 150-foot increments? Because the net fencing can shrink after a few years, and then it won't reach across the rectangles.

Suppose, for example, that you build a 5-strand permanent high tensile fence (more on this later) around a 580- by 1500-foot area. Then you can subdivide it lengthwise into four, 145- by 1500-foot smaller rectangles, using portable posts and three strands of polywire. (Step-in posts that have clips at correct positions are available. Multiple reel assemblies and single reel, multiple polywire combinations are also available for quickly and easily positioning three or more strands of polywire.) Then net fencing can be used to subdivide the 145-foot wide rectangles. Three sections of net fencing should be used for each group of animals: one section in front, another in back, and the third ahead of them to limit the next area that they will move into.

Keep records of when and where portable fencing was placed, and relocate the fencing at the same positions every rotation. This is to be certain that all areas of the pasture receive adequate recovery periods. If you don't relocate the fencing at the same positions in every rotation, some areas may get less recovery time than they need, and start you into untoward acceleration. When using portable front fencing, tie a short piece of twine or surveyors' tape to the perimeter fence at each spot where you connect the front wire(s). Relocating net fencing at the same positions is fairly easy, because of the clear checkerboard effect on the pasture that results from using this method for grazing sheep or goats.

LANES

Lanes should be as short and as direct as possible, to allow animals to reach all paddocks with the least amount of walking. Make lanes only as wide as needed, because forage in lanes usually becomes fouled and wasted. Lanes should be just wide enough to allow the animals and necessary machinery to get through. Animals will conform to lane widths, and walk single-file if they aren't rushed. For example, 12-foot wide lanes work well for animals, but aren't wide enough for large machinery.

Lanes 18- to 24-feet wide are needed for tractors with mowers, rakes, balers, and wagons.

Locate lanes on the highest and driest ground possible. Use culverts and gravel fill if necessary to make certain that your animals can walk through the lanes without dropping out of sight in mud. This is especially important in the earliest part of the season, when grazing must begin so that the forage doesn't become too mature, and during wet fall conditions.

The main idea in improving lanes is to shape them so that water drains off. Wisconsin dairy farmer Charlie Opitz has developed a way of stabilizing lanes that allows a lot of cows (800) to pass back and forth twice a day regardless of weather. First he grades the lane level to keep dump trucks level when spreading crushed stone. Then he spreads 15 yards of 1.5- to 2-inch crushed stone per 130 feet of 12-foot wide lane. On top of this he spreads 15 yards of ground limestone per 175 feet of lane. This currently costs about $2.00 per foot of lane.

Fabrics are now available that can be placed over muddy spots before covering with gravel. This keeps mud from mixing with the gravel, and forms a stable surface over wet areas in lanes and around drinking tanks. The fabric and gravel fill should be put in when the areas are dry, for best results.

An ideal lane and gate arrangement has gates in paddock division fences and at both paddock corners in V-shaped openings on the lane (Figures 6-1, 6-2). This design allows animals to move in any direction in the lane or to adjacent paddocks. If you can't build the ideal setup, be sure to place lane gates in the corner of paddocks nearest the barn and/or water. Otherwise, you may have some animals that can't figure out that they need to walk away from the barn to go through a gate into the lane before going to the barn. If this happens, instead of all of your cows coming to milk when called, you'll have to go to the paddock and chase out the few cows that get trapped in the paddock corner nearest the barn. This is especially a problem with heifers, and when you're just beginning to control grazing, because animals take a few days to learn what they're supposed to do.

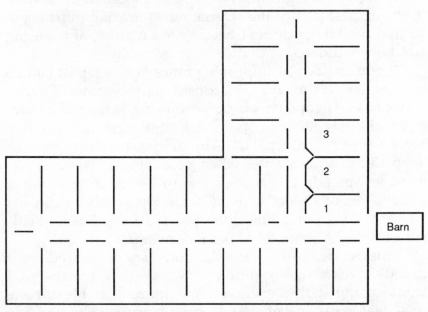

Figure 6-1. Example of a paddock and lane layout on level-to-rolling land, with gate openings allowing access to all paddocks and lane from all directions. Paddocks 1, 2, and 3 show V-shaped openings with gates closed. (Courtesy of Gallagher Springtight Power Fence, Tunbridge, Vermont)

If animals don't have to leave the pasture for drinking water or to be milked, you can put gates in paddock division fences, and move animals to fresh paddocks without entering the lane. The lane can then be used only once per rotation for returning animals to the first paddock to start the next rotation, and can be grazed as a paddock itself, since the forage in it wouldn't become fouled by animals moving through it.

140

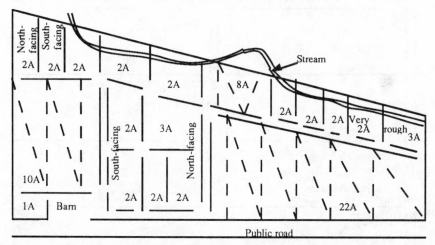

Figure 6-2. An example of paddock and lane layout on rolling-to-rough hilly permanent pastureland grazed by dairy cows. The 8-, 10-, and 22-acre paddocks are set aside in spring for machine harvesting. After adequate recovery time, these three large paddocks are grazed using portable fencing (polywire/post or tumblewheels) to subdivide them. Portable fence is moved every 12 hours, as indicated by --- lines. Back fence is installed in large paddocks after first two 12-hour grazings, and is then moved every 24 hours (See Figure 6-3 for back fence detail).

DAY/NIGHT PASTURES

If some of your pastureland for dairy cows is located a mile or more from the barn, it might be a good idea to use the land that's far away as day-pasture, and the closer land as night-pasture. This is so your cows don't take so long to come to the barn when called in the morning (you can call them well ahead of time for the afternoon milking). If you use day/night pastures, just be certain that recovery periods are adequate, and not too long, within the two areas. You can still use short (12-hour) occupation periods, but you may have to machine-harvest more forage in each area. This is because rotations within each area will take twice as long as they otherwise would have.

141

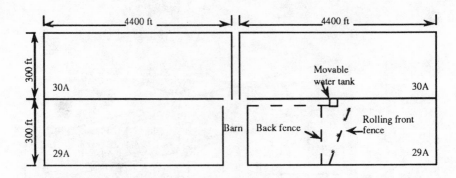

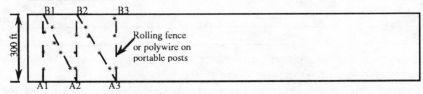

For cattle or horses:

First paddock is about 50 ft wide; for second paddock move A1 about 100 ft to A2; for third paddock move B1 about 100 ft to B2, and so on.

For sheep or goats:

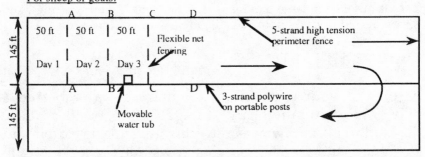

Move one section of flexible net fence once a day to give just enough forage for 24 hours. When animals are between sections B and C, move A to D, and so on, rotating around the large rectangles.

Figure 6-3. Examples of using large rectangles subdivided with portable fencing on level-to-rolling pastureland. (Not drawn to scale.)

DRINKING WATER

The source of drinking water for a pasture area is another

important consideration in planning a paddock layout. When one group of animals grazes a pasture, only one source of drinking water is needed, but they must be able to reach the water at all times. If possible, it's better to provide water in all paddocks, even though only one group may be grazing a pasture, because the animals need to walk less if water is nearby, and nutrients from manure and urine remain in paddocks where they're needed, not in lanes and around water tanks. When two groups of animals graze the same pasture area, water has to be provided in each paddock, since one of the groups must be kept in a paddock to prevent the two groups from mixing together.

With one group grazing and only one source of water, use lanes to allow animals to reach the water. Don't build triangular-shaped paddocks that radiate out around a central water supply, in regions having reliable, well distributed precipitation (nonbrittle) and moist soils (especially fine-textured) during the grazing season. This is a good design that allows easy handling of animals, but in the small paddocks usually needed on farms in humid regions it can result in transfer of nutrients from wider parts of paddocks to narrow parts near the water supply, and compaction of moist soil in narrow areas. This is because animals (especially dairy cows) prefer to be near their water supply, and will graze in the wide areas but camp in the narrow ones. Extremely high levels of nutrients can build up in the narrow areas, while nutrient deficiencies develop in wide parts. This nutrient transfer combined with soil compaction causes the pasture plant community to deteriorate. In arid regions, where soils are drier and paddocks are large, the advantages of having one water source at the center of radiating triangular-shaped paddocks outweigh any disadvantages.

Use 3/4-inch black plastic tubing to carry water to all paddocks. It's not necessary to bury the tubing; it can be laid on the soil surface in lane fence lines. Plants growing over it will shade it and prevent water in it from becoming too hot. In the fall the tubing can be drained by opening both ends, picking it up at one end, and walking to the other end while holding the tubing up and moving it hand over hand. Then it can be left on the soil surface during winter, without damaging it.

Water can be supplied to paddocks from the tubing in at least two ways:

1) Make connections to float valves on oblong-shaped, 40- to 100- gallon tanks for cattle and horses, and place one tank under every other paddock division fence. In this way two paddocks are served with each tank. Two- to 5- gallon tubs or buckets can be used in the same way for sheep, goats, and other small livestock.

2) Place garden-hose connectors in the tubing at intervals (e.g. every 300 feet) that will allow you to use a garden hose (e.g. 150-foot long) to connect to a full-flow float valve for one tub or a tank that you move along with the animals. Lightweight 40-gallon plastic or fiberglass tanks, valves, and connectors are now available from many dealers that sell fencing materials. The small tanks are convenient to empty and move by hand, without using skids and a tractor or truck. If there's good water pressure and you use a full-flow valve, a 40-gallon tank works well for 80 to 90 cows. Movable tanks are especially useful for providing water to animals grazing within portable fencing in large subdivided rectangles. It's best to move a 2- to 5- gallon tub along with small livestock every time they're moved within the net fences. With large livestock grazing within front and back fences, a tank can be moved when the back fence is moved, or less frequently by leaving it in wide spots formed in the back-fence lane.

A benefit of moving a water tank in the grazing area

with animals is that smaller tanks can be used, since all of the animals don't need or want to drink at the same time. If they have to wait and walk to water somewhere else as a herd or flock, much larger tanks are needed.

ELECTRIC FENCING

To manage grazing, livestock must be controlled. This means that you have to be able to place animals where you want them for as long as you want them there. Proper grazing management requires dependable fencing. Fortunately, having good fencing no longer is a problem.

Energizers

With the development of low-impedance electric fence energizers in New Zealand, controlling livestock became easier than it had been with ordinary fence chargers and nonelectric physical barrier fences. Energizers are effective because they produce an extremely short, high energy DC electrical pulse of about 6000 volts that charges fence wires once per second, or randomly every one to seven seconds. The short pulse brings fence lines up to high voltage so quickly that the energy isn't easily drained off, even if rain-soaked plants touch the fence wires, or a short length of wire has fallen on the soil. Also, the short pulse generates very little heat, so there's almost no danger of an energized fence causing a fire.

Energizers are available to fit all needs and circumstances. There are models that run on 120 or 220 volts, 12-volt auto or marine batteries, solar energy, or flashlight or lantern batteries. The different models have power outputs to charge fencing lengths ranging from 300 feet of net fence or 1,000 feet of polywire, to 4 weed-loaded miles or more than 25 miles of clean five-strand fence, and more than 100 miles of single-strand fence.

Energizers outperform all ordinary fence chargers in effective power by 2.5 to 20 times, depending on the size of the energizer. Although they dependably deliver a lot of power, energizers use very little electricity to produce it. For example, an energizer that will charge over 2 miles of five-strand fence or 60 miles of one-strand fence consumes only 8 watts of electricity!

Energizers cost somewhat more than ordinary chargers, but they're well worth the extra expense. Any extra costs are soon repaid in savings of time, labor, and materials needed to build and maintain an energized fencing system.

Because energizers don't short out easily, animals (livestock or predators) that touch the fence always receive a strong electrical shock that makes them respect the fence and avoid it. This dependable shocking power is why energized fences can be psychological or mental barriers, rather than physical barriers. Mental-barrier fences can be greatly simplified, and cost much less to build and/or maintain than physical-barrier fences, such as woven wire, multiple-strand barbwire or nonelectric smooth wire fences. (Never electrify barbwire, because animals or people can be cut badly by the muscle spasm resulting from the shock.) For example, a five-strand, spring-loaded high tensile fence for sheep, including the energizer, costs less than half as much per running foot than woven wire fencing.

Ordinary Chargers

An energizer makes it easier to control grazing, but it's not required, except for sheep. It's still possible to control cattle, horses, goats, and pigs with ordinary chargers. But because chargers are so easily shorted out by even one animal or wet plant touching the fence, they are not as dependable as energizers. An ordinary charger generally is one of the weakest links in a fencing system, and should be replaced with an energizer if you have any doubts about the charger's ability to control livestock. Animals may not respect a fence powered by an ordinary charger, and can easily knock down all your paddock division wires if the charger shorts out for even a little while.

If you have a good charger, its performance can be improved by:

1) Making certain that it is well grounded. Follow the instructions below for grounding energizers and chargers.

2) Keeping all fence lines clear of tall grass and weeds, since ordinarily charged wires can't touch anything but insulators. If only one or two wires are used with cattle or horses, the animals can graze underneath the wire(s), keeping fence lines free of tall weeds and grass. Thistles and other unpalatable plants still will have to be cut, but the amount of time and effort needed for clearing fence lines will be less.

3) Making certain all insulators are in good condition, and that wires don't touch posts, trees, or brush.

4) Cleaning fence lines of all old wire. An old barb wire fence running through weeds and brush and nailed to trees makes a good ground. If electric fence wires touch the barb wire, chargers and energizers will short out.

Supposedly, sheep can be kept in with ordinary fence chargers, but not easily, because their wool insulates them from the electrical pulse produced by ordinary chargers. Keeping sheep in with an ordinary charger requires a very good charge, short fence, and few or no plants touching the wires. In my experience, once lambs or ewes learn how to go through a fence having low voltage, it's nearly impossible to keep them from doing it again, unless fence voltage is brought to at least 5000 volts. Ordinary chargers can't be used with the five-strand perimeter fencing needed to keep predators away from sheep or goats, because the lower wires must be close to the soil where a lot of plants can touch them. Net fences for small livestock also touch plants or soil, which would easily short-out an ordinary charger.

Grounding Energizers and Ordinary Chargers

Energizers can fully charge very heavy fence loads and discourage even the most unruly livestock from going through fences. But to do this, they must be well grounded. The performance of ordinary chargers also can be greatly improved by making certain that they are well grounded. Instructions for making a good ground that usually accompany energizers are inadequate for many soils, especially those formed from glacial till.

Since the ground is half of the electric fence circuit, and grounding instructions with ordinary chargers are totally inadequate for all locations, I think that it's worthwhile to explain here how to make a good ground:

1) Drive four to six grounding rods (galvanized pipes or copper rods) 40 to 50 feet apart, preferably into moist soil, beginning near the energizer or charger. The grounding rods should be 6- to 8-feet long, and should be driven straight down, if possible, to be in moist soil. If they can't be driven straight down, because of ledge or rocks, drive them in at an angle. Leave 4 inches of rod sticking out above the soil surface, to make connections.

The rods can be placed in any formation, just so they're out of the way. If possible, place them so that one connection in the grounding system can be made to a metal culvert under a road or lane, or to an unused well casing. The more powerful the energizer or charger, the more grounding rods are needed. In dry sandy or glacial-till soils where grounding is difficult, 10 rods or more may be needed, if metal culverts aren't available to connect to.

148

2) Clean connection points on the rods with a file, wire brush, or steel wool. It's best to use one length of galvanized wire from the energizer or charger ground terminal to all grounding rods, to make certain that the connection is continuous. Double the wire for grounding larger energizers. Attach the wire securely to each rod with stainless steel clamps or galvanized bolts.

3) Test the grounding system. Before plugging the energizer or charger into its power source, connect the unit to the fence and grounding system. Then load the fence by laying about 200 feet of fence wire on the soil, several hundred feet away from the energizer or charger, or attach the electric fence wire to a metal object such as a long, old woven-wire fence. Then plug the unit in and test the grounding system with a digital fence tester, by taking a reading between a grounding rod and soil about 4 feet away. You can also do this test by using your two hands spread well apart, touching the soil and the ground wire..... If the grounding system is charged, improve the ground until it carries less than 400 volts when tested. If you hadn't connected to a culvert already, the easiest way to improve a ground is to connect it with bare wire to the nearest metal culvert, even if it's several hundred feet away from the energizer or charger.

Lightning Arrestor and Choke

Energizers and chargers must be protected from lightning, which could induce high voltage on a fence line that would damage the unit and possibly start a fire in the building where the unit is kept. Lightning arrestors and a choke (induction coil) installed on a fence line can protect energizers or chargers against lightning-induced voltage.

At least one lightning arrestor should be placed at the point where the hot wire from the energizer or charger attaches to the fence. For more protection, other arrestors should be installed at each corner of the fence, or one per mile of fence. In areas that have a lot of lightning, a choke should be installed between the energizer or charger and a lightning arrestor. The choke creates resistance for the extremely high voltage of lightning, causing it to go through the lightning arrestor and into the earth. For total protection during electrical storms, the energizer or charger should be unplugged, with the fence and ground terminals disconnected.

The ground for lightning arrestors can't be part of any other grounding system. Since lightning always takes the shortest and easiest path to the earth, the ground for lightning arrestors should be as good as the grounding system for the energizer or charger. If the grounding system for the energizer or charger is poor, arrestors won't protect the units from damage.

Besides protecting your energizer or charger from lightning-induced high voltage, it's also a good idea to have an electrician install a large surge suppressor in your electrical supply. This will protect your unit from high voltage surges coming in from the power line.

Training Animals

To avoid discouraging problems, animals should be trained to respect an electric fence before turning them out to pasture. To do this, build either a strong multiple-strand electric fence or a wooden rail fence with one or two strands of electric wire toward the inside, around a barnyard or similar small area. An obvious barrier, such as wooden rails, on the other side of the electric wires trains animals to back away when they get shocked. Place some grain on the soil at various spots just inside of the electric wires. Connect the energizer or charger to the electric training fence. Confine animals to the area for several hours (always have water available to them), until you're certain that at least some of them have been shocked and they all respect the fence.

FENCING MATERIALS AND CONSTRUCTION

We are fortunate that New Zealanders didn't give up on pasturing livestock as others did, but continued to develop fencing materials that would enable them to do a better job of managing grazing. Because of this and recent developments in the United States, New Zealand energizer and fencing technology has been combined with Yankee ingenuity to make available extremely effective electric fencing. These fencing materials are relatively inexpensive, easy to build and maintain, and long-lasting. Let's look at some of the materials and construction techniques that are available. (Excellent detailed instructions on building high tension electric fences are available from dealers of fencing materials, including the companies listed below.)

High Tensile Wire

High tensile smooth fence wire is much stronger (1,000-pound breaking strength) than common soft smooth wire. This enables a high tensile fence to withstand a charging animal or winter snow loads without breaking. High tensile wire is elastic and flexible, but doesn't stretch like soft wire does. It has a heavier galvanized coating that makes it last at least three times longer than common soft galvanized smooth wire. (Special soft wire is now available that is heavily galvanized to make it last longer.)

Springs and Tighteners

In fence runs of less than 450 feet, the elasticity of high

151

tensile wires maintains tension on the fence. In fences over 450 feet long, a spring installed in each 1500 feet of wire maintains about 150 pounds of tension on the wire. In areas where there are snow and icing conditions, fence runs shorter than 450 feet should also have springs to prevent the wires from breaking. Tighteners placed in each wire near springs are used to take up slack in the wire and tighten it to the desired tension.

Posts

Corner posts must be strong, firmly set in the ground, and well-braced or anchored to hold high-tensioned wires. Corner posts should be at least 6 feet long, and can be 3- x 3-inch sawn hardwood, or 6-inch diameter softwood (usually cedar or treated pine). Trees conveniently located at corners can also be used with J-bolts and insulators to hold wires.

Line posts serve only to hold the wires up and maintain proper wire spacing. So line posts can be only 5 feet long, fairly light (1 5/8 x 1 5/8 inches), and can be spaced up to 150 feet apart. When line posts are widely spaced, lighter (1 1/4 x 1 1/4 inches) battens that stand on top of the ground are used to maintain proper wire spacing. Battens and anchors are also used to hold the wires down where fences cross low spots in uneven terrain. Battens are needed about every 35 feet for five-strand fences, and every 50 feet for two- or three- strand fences. Four-foot, 1 1/4- x 1 1/4- inch line posts are used instead of battens for keeping one-strand fences in place.

Posts made from dry hardwood that has been pressure treated with creosote are self-insulating, and no insulators are needed in attaching wires to these kinds of posts. Hardwood line posts have sharpened points so that they can be easily driven into the ground with a post driver or light sledge hammer. Line posts and battens are slotted to hold wires at proper spacings. Wires are held loosely in the slots with heavy wire clips. This allows fence wires to move in the slots, and keeps the fence flexible so that it can withstand stress, and spring back into place without breaking.

Wire Spacing

High-tension perimeter fence wire spacing for different livestock should be as follows for:
1) Dairy, beef, horses:
One strand for cows, yearling heifers, and horses: 28 to 30 inches from the soil surface; 46-inch fences are needed for some horses. (Low-tension single-strand smooth wire or polywire, also can be used with portable posts at 28 to 30 inches from the soil surface to make internal subdivisions.)
Two strands for cows with calves, 6- to 12-month-old heifers, and horses with foals: 17 and 37 inches from the soil.

2) Sheep and goats:
Five-strand perimeter fence for keeping predators out: 6, 11, 17, 26, and 37 inches from the soil surface.
Three-strand wire (or polywire portable fencing) for internal subdivision of large paddocks: 7 to 9, 17, and 28 to 30 inches from the soil.

3) Pigs:
Two strands: 7 and 17 inches from the soil, or:
Three strands: at 6, 13, and 23 inches.

Insulators

Insulators must be used wherever anything but dry

153

treated hardwood posts are used. Insulators must be strong enough to support long spans of high-tension wire, and large enough to allow wires to slide freely.

Cutout Switches

Paddock layouts should be designed so that separate sections can be shut off with DC cutout switches. Convenient switch locations are at gates, corners, and places where one large pasture area ends and another begins. The switches allow you to shut off unused sections of paddocks and put full electrical power into the section(s) containing animals. In case of a problem, having paddocks in separate sections allows you to test each section in sequence, starting with the one nearest the energizer or charger.

Portable Fencing

Lightweight portable fencing is easily moved and reused at many locations during a grazing season. It costs more per foot than permanent fencing, but much less is needed. Since it takes the place of large amounts of permanent fence, portable fencing can save a lot of money. Portable fencing can be used with energizers and solid-state ordinary chargers, but not with "weed-burners". The long pulse of "weed-burners" creates heat in fence wires, which would melt plastic portable fencing.

Polywire

Polywire is a polyethylene twine braided together with six strands of overlapping stainless steel to carry the electric charge. It usually comes in 650- or 1600-foot rolls. Polywire has many uses, including subdividing large permanently fenced paddocks, temporarily fencing around irregular areas or hill contours, making gates, and looping down over stream beds to keep large animals from walking beneath permanent paddock division wires. When used to subdivide large paddocks, polywire is supported on plastic or fiberglass posts, or on tumblewheels. Polywire shouldn't be used as a mainline fence to power other fences, because it doesn't conduct electricity well enough.

Polytape

Polytape is a polyethylene tape 0.5 to 1.5 inch wide that has 5 to 11 strands of stainless steel embedded in it. Polytape comes in 650 or 1600-foot rolls, and is used for the same purposes as polywire, but is more easily seen by livestock than polywire. This may be especially helpful with some animals that can't see small-diameter wire very well, or that are especially afraid of electric fence, such as horses. Polytape is very useful as a training fence.

Reels

Completely insulated hand reels are used to roll up polywire or polytape. The reels are light, well balanced for handling ease, and can be hooked onto perimeter fences. It's possible to attach three to five reels to a short steel post, that then allows three to five polywires or tapes to be positioned at the same time for internal fencing of small livestock. It's also possible to have three shorter separate strands of polywire on one reel, and position the three polywires at the same time.

Rolling Fence Wheels

Rolling fence wheels (e.g. Gallagher tumblewheels) essentially are six-spoked, rimless rolling fence posts, that

155

hold up and are held up by polywire or polytape ahead of and behind grazing animals. They're quickly and easily moved, inexpensive, and made of very light noncorrosive materials.

Electric Net Fencing

Electric net fencing for small livestock consists of six horizontal strands of polywire woven to form a large-mesh net, with vertical braided or solid strands of nonconducting polyethylene. About every 12.5 feet a round plastic post is incorporated into the netting to hold the fence up.

Plastic and Fiberglass Posts

Plastic posts with wire holders molded into the posts at proper wire-spacing intervals are available to make installing multiple-strand portable fencing very easy. They also have a large pointed rod on the bottom that goes into the soil, and a lip to step on for pushing them in. This means that you can carry a bundle of 20 or more posts, pick one from the bundle, and step it into the soil without setting the bundle of posts down.

Round fiberglass posts with adjustable plastic wire holders are useful for making single-strand polywire or metal wire fences. Round and T-shaped fiberglass posts are available, but the round ones are better made and easier to use. The T-shaped posts give off a lot of small splinters that are very painful and nearly impossible to find and remove. Round fiberglass posts also give off some splinters, so it's a good idea to wear gloves when handling any fiberglass posts. Also, wires must be held to the T-shape posts with metal clips that don't allow the wires to move, so they can't be tightened without removing all of the clips. Another problem is that every time an animal hits the fence hard enough, a lot of clips pop off the T-posts; the clips could end up inside animals.

Physical-Barrier Fencing

If you already have good physical-barrier fencing, by all means use it. I don't want you to get the impression from the above discussion that to control grazing you have to spend a lot of money building new fences. Quite the contrary: spend as little as possible getting started. This book is to help you to manage your pastures in a way that will make your farm more profitable. I certainly don't want you to go further into debt in setting up this system. Because electric fencing costs less than physical-barrier fencing, it's usually best to replace physical-barrier fence as it wears out with long-lasting, low-maintenance high tension electric fence.

Physical-barrier fencing works perfectly well in most situations, especially as perimeter fencing for large livestock. Just make certain that the fencing is tight and in good condition. When used for paddock divisions, however, barbwire and nonelectric smooth wire don't work very well, because animals gradually loosen it up by reaching underneath and through it to eat forage in adjacent paddocks. It's best to use portable or low tension electric fencing to subdivide areas within good physical-barrier perimeter fencing.

The useful life of physical-barrier fencing can be extended by attaching special offset brackets with insulators to the fences, and adding an electrified smooth wire to them. This prevents animals from pushing against and loosening the fences. This offset electrified wire can also be used to carry power to portable or permanent subdivision fences. Don't attempt to run an electrified strand along a nonelectric fence that is in poor

157

condition: it will cause you more problems than it's worth, by shorting out your electric fence system.

In some countries neither energizers nor ordinary chargers are available, or they're too expensive for farmers to buy because of import taxes. Although electric fencing makes it easier to subdivide pastures, physical-barrier fencing certainly can be used to build paddocks. Rock walls, split rails, stumps and brush, bamboo, living briars and hedges, woven wire, and multiple-strand barbwire or smooth wire are examples of materials useful for making physical-barrier fences. It doesn't matter how pastures are subdivided, just so they're divided enough to provide adequate recovery periods for the plants under local conditions, and short occupation periods.

SOURCES OF ENERGIZERS & FENCING MATERIALS

Your local University Extension Service and Soil Conservation Service should have information about dealers that sell energizers and fencing materials, and build high tension electric fencing systems. But if they don't have the information, this list of companies should help you to find a nearby dealer:

Gallagher Power Fence, Inc., P.O. Box 708900, San Antonio, TX 78270. Phone: 800/531-5908 (512/494-5211 in Texas).

Gallagher/Spring-Tight Power Fence, RFD 1, Box 158, Tunbridge, VT 05077. Phone: 800/872-9482 (889-3737 in Vermont).

Grassland Supply, Rt. 3 Box 6, Council Grove, KS 66846. Phone: 800/527-5487.

Jeffers, P.O. Box 948, West Plains, MO 65775 and P.O. Box 100, Dothan, AL 36302. Phone: 800/533-3377.

Kencove, 111 Kendall Lane, Blairsville, PA 15717. Phone: 412/459-8991.

Kentucky Graziers Supply, 1929 S. Main Street, Paris, KY 40361. Phone: 606/987-0215.

Kiwi Fence Systems, Inc., RD 2 Box 51A, Waynesburg, PA 15370. Phone: 412/627-8158.

McBee Agri Supply, Inc., Rt. 1 Box 121, Sturgeon, MO 65284. Phone: 314/696-2517.

Perfect Pastures, RR 1, Highland, WI 53543. Phone: 800/829-5459.

Premier Fence Systems, P.O. Box 89, Washington, IA 52353. Phone: 319/653-6631.

Tipper Tie, P.O. Box 866, Lufkin Road, Apex, NC 27502. Phone: 919/362-8811.

Trident Enterprises, Inc., 9735 Bethel Road, Frederick, MD 21701. Phone: 301/694-6072.

Twin Mountain Supply, P.O. Box 2240, San Angelo, TX 76902. Phone: 800/331-0044 (800/527-0990 in Texas).

West Virginia Fence Corporation, U.S. Route 219, Lindside, WV 24951. Phone: 304/753-4387.

CHAPTER 6

REFERENCES

Gallagher Electronics Ltd. 1985. *Gallagher Insultimber Power Fencing Manual.* Hamilton, New Zealand. Brochure. 24 p.

Gallagher Electronics Ltd. 1985. *Gallagher Temporary Power Fencing Systems.* Hamilton, New Zealand. Brochure. 6 p.

Gallagher Electronics Ltd. 1985. *Tumblewheel: the Simple and Effective Time Saving Aid to Strip Grazing.* Hamilton, New Zealand. Brochure. 2 p.

Patenaude, D. and J. Patenaude. 1991.*The Grazing News.* Perfect Pastures, Rt 1, Highland WI. Newsletter. March.

Pel Electric Fence Systems. *Instruction Manual.* No address or publisher given. 28 p.

Premier Fence Systems. *The New Fencing Systems Made Simple: A Do-It-Yourself Guide to Buying and Building Better Fences.* Premier, Box 89, Washington, Iowa. 39 p.

Swayze, H.S. 1984. *Gallagher Spring-Tight Power Fence (Construction, Materials, and Costs).* Tunbridge, Vermont. Brochure. 8 p.

Livestock Production

Miraculous power and marvelous activity --
Drawing water and hewing wood!

Pan-yun

Animal evolution has been greatly influenced by diet available, so animal and plant development have been closely linked. Since plants contain cellulose, which is a polymer of glucose that makes up part of the fiber in plants, it's interesting that animals didn't evolve enzymes capable of breaking down cellulose. It's especially surprising because cellulose is one of the most abundant naturally occurring organic compounds that can be used for food.

The grazing animal's life style may have had a lot to do with how things turned out. The defenseless grazers probably would leave the forest cover only to fill up quickly on plants growing in clearings near the forest edge. Once full, they would return to hiding places to digest the food. Instead of cellulase enzymes, grazing animals developed digestive tracts in which vast numbers of symbiotic microorganisms live and digest fibrous and soluble plant materials. These microorganisms effectively became the animal's cellulase-producing tissue.

This kind of digestive symbiotic system reached its fullest development in ruminants (e.g. cattle, sheep, goats) which contain a large fermentation vat called the rumen. Ruminants use pasture forage by first swallowing it into the rumen for fermentation. Large coarse forage material (cud) is regurgitated, rechewed, and swallowed into the rumen again. Most of the soluble nutrients are consumed rapidly by rumen microorganisms.

Some unfermented fiber does leave the rumen, but it and other food material and microorganisms that pass out of the rumen are worked on by normal enzymatic digestion in the lower gastrointestinal tract. The rumen absorbs nutrients produced by microbial fermentation, such as acetic, propionic, and butyric acids, which contribute 60 to 75 percent of the animal's energy requirements. Rumen microorganisms also produce amino acids that are essential to ruminants for making proteins. Vitamin K and the B-complex vitamins are all synthesized by microorganisms in the rumen.

Nonruminants (e.g. horses, pigs) have much smaller digestive tracts than ruminants, and feed consequently passes through them faster. Feed passes through a horse's stomach in about 24 hours, whereas feed remains in a cow's rumen for about 72 hours. Compared to ruminants, relatively little microbial action occurs in nonruminant stomachs, and what does occur is located in the cecum, after the small intestine. This contrasts with the rumen, which is located before the small intestine. For this reason, ruminants have an advantage of greater absorptive capacity compared to nonruminants.

Feed digestion in poultry (e.g. chickens, turkeys, ducks, geese) occurs in two organs: the proventriculus and the gizzard. Hydrochloric acid and enzymes are secreted into feed before it passes to the gizzard, where it's ground and mixed with digestive fluids.

What all this means simply is that nonruminants and poultry have different nutritional requirements and must eat more frequently than ruminants. Despite differences, most or all of the nutritional needs of all livestock can be met on properly managed pasture.

To estimate potential livestock productivity from pasture, we must know three things: 1) nutritive value and yield of the pasture forage; 2) feed requirements of grazing animals for a particular level of performance (e.g. rate of weight gain or milk production) within the physiological states of maintenance, growth and fattening, or lactation; and 3) total animal need and distribution of need for forage during the grazing season.

PASTURE NUTRITIVE VALUE AND LIVESTOCK MINERAL REQUIREMENTS

Protein, energy, and fiber contents of pasture forage determined by laboratory analyses are indicators of its nutritive value. High-quality forage contains low amounts of fiber and high levels of protein and energy; in this respect legumes at the same stage of growth are better than grasses. Season and stage of growth within a season affect the nutritive value of pasture forage. Protein, energy, and fiber values of pasture plants are most conducive to animal production when the plants are grazed at young stages of growth.

In properly managed pastures crude protein content of the forage can average more than 22 percent over a 6-month grazing season. Based on National Research Council (NRC) standards, this qualifies well managed pastures to be defined as protein supplements! (Protein supplements are feedstuffs which contain more than 20 percent crude protein.) Forage from excellent pastures can contain as much or more energy as in barley, oats, triticale, or wheat. This is enough to meet the energy needs for all but the highest producing animals at peak production, depending on intake. The dry matter content of forage of well managed pastures ranges between 20 and 25 percent.

Grazing management indirectly affects pasture nutritive value by controlling the plants' stage of growth. Short (3-4") leafy pasture is typical of good feed for sheep when they are turned into a paddock. Longer (5-6") leafy plants characterize pasture for cattle when it's ready to be grazed. Preflowering and flowering stages correspond to forage ready for haying or ensiling.

When grasses grow past the head-emergence stage, digestibility decreases. Periodic grazing, however, returns pasture plants to physiologically young stages of growth. Because of the effects of temperature and daylength on plant growth, the leaf:stem ratio, protein, energy, and fiber contents still change in plants that have been grazed. But the decrease in digestibility and increase in fiber of grazed plants are very much less than in plants allowed to grow uncut or ungrazed.

Besides protein, energy, and fiber, pasture forage also contains low amounts of fat, and variable amounts of minerals and vitamins, depending on soil fertility and plant growth stage. Mineral needs of livestock vary widely. The best ways to make certain that adequate levels of minerals are present in pasture forage is to feed the soil by applying manure, lime, and fertilizer, and maintain as complex a plant mixture in the pasture as possible. No one species of plant is good at everything, and one plant species can't satisfy all of the mineral and other nutritional needs of grazing livestock. Green forage from well managed pastures with mixed plant communities usually contains all the vitamins needed by livestock.

Salt and Trace Elements

Sodium and chlorine are needed by all animals, especially plant eaters. Since forage plants may not contain enough of these elements to satisfy animal needs, salt (sodium chloride) should be available at all times to animals on pasture. It should be provided free-choice as loose granules rather than as blocks, because animals may not get enough salt from blocks to meet their needs, especially during hot dry weather.

If there's any doubt about the availability of trace elements such as cobalt, copper, iodine, manganese, selenium, or zinc in your pasture soils, provide a mineral mix or salt that contains the needed trace elements. Trace element supplementation may be especially critical if you aren't feeding grain concentrates, which generally contain added trace elements. In some situations the greatest response to concentrate feeding may be due to trace elements contained in the concentrate, rather than to the protein and/or energy they contain. It's usually a lot less expensive to provide needed trace elements with or without salt, rather than in concentrates. Be sure to provide trace element mixtures suitable to the kind of livestock you're raising and your local conditions.

Carefully place the salt and mineral supply at a location in each paddock where you want the animals to spend most of their time in that rotation. By placing salt and minerals at different locations in each paddock in every rotation, you can distribute the animals and their manure more uniformly throughout the paddocks, and avoid overgrazing areas around the salt and minerals. Placing the salt and mineral supply in the middle of an area containing unpalatable weeds helps to destroy the weeds by trampling.

The amounts of salt needed in pounds per animal per

month are: 3 to 5 for mature cattle, 1 for mature ewes, 1.5 for milking goats, and 5 pounds for horses. The salt requirement of horses increases greatly when they work hard and sweat a lot. Salt is also one of the minerals needed in largest amounts by pigs, and should be available to them free-choice at all times. Poultry require salt at the rate of 0.2 to 0.5 percent of their diet, which they can obtain if loose salt is made available. Lactating animals need more salt than dry ones. Young animals need less salt than mature ones.

Calcium and Phosphorus

Next to sodium and chlorine, grazing animals are more likely to experience a deficiency of phosphorus than any other element. Although most forage plants contain adequate amounts of calcium, they may be low in phosphorus, especially if the phosphorus level in the soil is low. Legumes are excellent sources of calcium, but grasses contain less calcium. Forage analyses can easily tell you how much calcium and phosphorus your pasture forage contains. Feeding the soil by applying rock phosphate or phosphorus fertilizers according to soil-test recommendations, usually increases forage phosphorus level and yield of pasture plants. If it's not possible to apply all the phosphorus that's needed because of the expense involved, or if phosphorus fertilization fails to raise phosphorus content of the plants to adequate levels, phosphorus supplements should be made available free-choice to the animals.

Calcium and phosphorus use by grazing animals depends on: 1) an adequate supply of available forms of both elements, 2) a suitable ratio between them, and 3) adequate Vitamin D from sunlight or the ration to enable calcium and phosphorus to be assimilated and used. If there's plenty of Vitamin D available, the ratio of calcium to phosphorus is less important. If the ratio of calcium to phosphorus is satisfactory, less Vitamin D is needed.

Calcium to phosphorus ratios in the total ration for ruminants should be in the range of 2:1 to 4:1, except for dry cows, which need a ratio of 1.2:1 to 1.7:1. Usually the calcium to phosphorus ratios for horses should be about 1.1:1. Growing pigs and poultry require a ratio of 1.2:1 to 1.5:1. It's always important to have more calcium in feed than phosphorus, but not too much more. As production increases, needs of calcium and phosphorus (and all other minerals) also increase.

If calcium or phosphorus supplements are needed, the amount and source used depends on the mineral composition of the total ration. If only calcium needs to be supplemented, provide ground limestone or oyster shells. If only phosphorus is needed, provide monosodium phosphate, disodium phosphate, sodium tripolyphosphate, or feed-grade phosphoric acid. If both calcium and phosphorus are needed, provide dicalcium phosphate, steamed bone meal, defluorinated rock phosphate, or a commercial mineral mix. Provide the minerals freechoice, mixed with about 40 percent salt to ensure that adequate amounts of the supplements are eaten. Crushed oyster shells should always be provided for grazing poultry to satisfy any need for extra calcium.

Milk Fever

Dry cows are a special group that should be managed separately during the last month of their dry period, to avoid having them come down with milk fever when they freshen. Milk fever results from a lowered blood calcium level around the time of calving. A cow has large stores of calcium in her skeleton and plenty in the feed in

167

her digestive tract. But she only has a small amount circulating in her blood, and this isn't enough to meet the drastic change from the needs of the developing calf, to the demand of milk production in early lactation. The deficiency is only temporary, however, because the cow has ample reserves of calcium. A cow suffering from milk fever can't mobilize calcium reserves quickly enough. A few days after calving the rate of calcium mobilization from reserves increases, and she is then able to cope with the demands of lactation.

Calcium flows from the blood into the milk, and unless blood levels are replenished from reserves in bones and intestines quickly enough, blood calcium level falls, and milk fever results. If untreated, muscle paralysis occurs and the cow rolls over on her side, and is unable to get up again. When she is on her side, rumen gases can't escape, so she becomes bloated and dies from either pressure on her heart or from inhaling rumen contents. Treatment requires a timely injection of calcium and care by a veterinarian.

Milk fever mainly occurs in second-calf heifers and older cows. Older cows are much more susceptible because the calcium reserves in their skeletons are less available. Jerseys are more susceptible than other breeds, because of the high demands of lactation in relation to their size and total metabolism. Milk fever usually isn't a problem with first-calf heifers.

Spring- or autumn-freshening cows grazing lush pasture may come down with milk fever, unless managed appropriately. Lush well managed pasture is not good feed for dry cows during the last month (especially last 3 weeks) of their dry periods, because calcium content of the forage is too high and fiber content is too low for their needs. During the last month of their dry periods, cows should be fed a low calcium, high fiber ration; their calcium mobilization process will then be active enough at calving to cope with the increased demand.

During the last month of the dry period, rations should have the following characteristics: 12% crude protein, 35% acid detergent fiber, 0.5% calcium, 0.4% phosphorus, and 0.60 Mcal/lb net energy lactation. They should be fed long stem dry hay to keep their rumens full and stimulate good rumen muscle tone. A few days after calving, fresh cows can go back on pasture. (G.M. Catlin, V.M.D. and J.R. Kunkle, V.M.D: personal communication)

Magnesium

The most common problem of magnesium deficiency in grazing animals is called grass tetany, grass staggers, or hypomagnesaemia. Cattle are more susceptible to grass tetany than sheep or goats, and females are most susceptible, especially when they're milking or pregnant.

Grass tetany occurs most frequently in animals grazing lush grass-dominated leys, highly fertilized with nitrogen and potassium. If soil pH level is low, plant uptake of magnesium is reduced even more. Grass tetany is rare or nonexistent in animals grazing pastures growing on soils that developed from dolomite. It occurs more commonly in the spring among unsheltered animals during cold stormy weather, especially within 2 weeks to 2 months after giving birth. Animals grazing pastures that contain legumes and other broadleaf plants (e.g. dandelion, plantain, chicory), are less likely to come down with grass tetany, because broadleaf plants contain more magnesium than grasses.

Grass tetany results from a low level of magnesium in the blood, which may be due to eating forage containing less than 0.2 percent magnesium. An imbalance of ions occurring when the ratio of potassium and sodium to

169

calcium and magnesium is unfavorable in forage may indirectly result in low magnesium intake and absorption.

To prevent grass tetany, pasture fertilization should be balanced by applying magnesium along with other nutrients. Manage pastures so that clover growth is encouraged. Also, animals should be given a magnesium supplement, especially during the spring in areas where grass tetany is a problem.

Animals affected with grass tetany become nervous, and develop leg muscle spasms and convulsions. Animals usually recover if treated with a timely injection of magnesium gluconate and sedatives.

Water

Water is the first limiting nutrient for all animals, and good-quality clean water must be available at all times to grazing livestock. Clean water tanks or tubs regularly so that the water doesn't carry dirt or pathogenic bacteria. A dirty water tank easily spreads parasites and diseases through a herd or flock. Water needs vary depending on size of animal, milk production, activity, dry matter intake, rate of gain, pregnancy, air temperature, and relative humidity.

All animals drink less water when grazing lush pasture forage that contains 75 to 80 percent water. For example, sheep may go for days without drinking when they're grazing lush pasture. But even if animals don't drink, keep water available to them.

Daily water needs (including water in forage) in gallons per animal per day are: 12 to 36 for milking cows, 5 to 20 for full-grown beef cattle, 1 for mature sheep, 0.5 for weaned lambs, 5 to 6 for milking goats, 10 to 12 for horses, and 0.5 to 1.5 for pigs. Poultry also must have free access to fresh, clean water at all times. Of course, ducks and geese like water for bathing and playing in, besides drinking.

PASTURE EVALUATION

Sampling

To evaluate pasture forage nutritive or feeding value and dry matter intake, representative samples of the forage must be taken and analyzed. Unfortunately, this is easier said than done. Entire books exist on techniques of sampling pasture forage, and none of the methods is completely satisfactory. The problem is that the analytical results are only as good as the sample. The difficulty lies in taking samples that represent the general nutritional state of forage present in the pasture, and the forage that animals actually graze.

Forage Feeding Value

We're using a pasture forage sampling method for nutrient analysis that's simple, inexpensive, and gives good results. It involves walking in a zig-zag path across a paddock just before it is grazed, and taking 30 to 40 subsamples of the forage. For each subsample take the amount of forage that you can hold and pull off with your thumb and two fingers, taking only what animals are likely to graze. Place the subsamples in a plastic bag as you collect them, combining them all into one composite sample that represents the overall forage in the paddock.

Sample only from areas that will be grazed, not in rejected forage spots around manure. Rejected forage around manure tends to be higher in phosphorus and potassium, and grows about twice as fast as forage of grazed areas. Rejected forage also becomes more mature and decreases in feeding value as the season progresses.

Since the forage is rejected, there's not much point in sampling and analyzing it.

Label the samples and freeze them until you can get them analyzed. Have samples analyzed as soon as possible after taking them, so that you know the quality of forage that your animals are eating and can adjust your in-barn feeding program accordingly. Using overnight shipping and FAX services, it's possible to have analysis results within 2 days of sampling. You can get a good estimate of the quality of forage available by sampling one or two representative paddocks every two weeks.

Forage Dry Matter Intake

Variation in the intake of pasture forage by grazing livestock has a major influence on animal performance. Sward botanical composition, leaf-to-stem ratio, pre- and postgrazing pasture mass, and forage allowance probably have more important effects on forage intake than nutritional aspects. The goal of grazing management is to influence these aspects to make available forage that's easily grazed, so intake can be as high as possible.

Estimates of dry matter intake of grazing livestock can be made for individuals using direct or indirect animal techniques, or for groups by pasture sampling. All estimates are just that; they aren't precise, no matter who made them or where they came from. All techniques used in obtaining estimates are subject to error due to variation in pasture swards, animals, and sampling techniques. Use estimates simply as guides to help in managing grazing.

The most common animal technique involves an indirect estimate of intake from laboratory determinations of pasture forage digestibility, and indirect or direct measurements of poop output of grazing animals. By this method: DM intake = poop output/1-digestibility.

Average dry matter intake for groups of livestock is estimated from differences between pre- and postgrazing pasture masses in paddocks of known areas. This method is best used over short occupation periods of less than 4 to 5 days. Intakes estimated by pasture sampling generally are 30 to 40 percent lower than intakes estimated by animal techniques. By this method:

$$\text{DM intake} = \frac{\text{pregrazing mass - postgrazing mass}}{\text{number of animals grazing x days}}$$

In the tables at the end of Chapter 8, estimated dry matter intakes appear to be high, especially for dairy cows. But we don't know yet how much forage livestock actually can or will eat when grazing well managed pasture in the United States. The values for intake (and protein and energy requirements) in the tables are based on New Zealand estimates, which were obtained from animals grazing pasture. This contrasts with values available in the United States that were obtained with stall-fed animals. A lot of assumptions have to be made when using stall feeding to get information about grazing animals. Not all assumptions are valid.

Protein

The amount of protein needed, if any, in a concentrate supplement depends on the quality and kind of forage that animals are eating in the pasture, and on the kind of animal, its physiological state, and its production level. In general, as protein content of forage increases, less protein is needed in the concentrate. The overall ration (forage plus grain and protein and energy supplements) should contain at least 14 percent total protein, except for ewes and beef cows, which don't need as much. High-producing dairy cows may need about 17 percent total protein to digest the energy in the extra feed that they require, and to make milk. (Further complications of protein feeding of dairy cows are discussed below.)

Under good feeding conditions of well managed pastures on fertile soils, grazing animals usually consume more protein than they need. For example, pastures under controlled grazing management in Vermont have averaged more than 22 percent crude protein in the forage over a 6-month grazing season -- more than enough for most livestock, if forage intake is adequate. If protein needs can be met with pasture forage, a lot of money can be saved in feeding costs, because protein supplements usually are very expensive.

Protein for Milking Dairy Cows

Nitrogenous components of various kinds, including true protein and nonprotein nitrogen, are used in the protein metabolism of ruminants. True protein components can be divided into fractions that are either degraded in the rumen, or undegraded and pass through the rumen (bypass protein). Most of the nonprotein nitrogen and rumen degradable fractions are broken down and used by rumen microorganisms in forming microbial protein. Generally microbial protein is used to meet protein requirements of ruminants, but it may not be adequate to satisfy the needs of dairy cows producing more than 35 (Jerseys) to 50 (Holsteins) pounds of milk per day. Protein that passes through the rumen undegraded can serve as a source of amino acids that can in effect supplement microbial protein to improve milk production.

Most of the protein and nonprotein nitrogen in high quality, lush pasture forage degrades in the rumen. Since the level of undegradable intake protein needed for

174

milking cows is thought to be about 37 percent of total protein, there may be special problems (or opportunities) in balancing rations for dairy cows being fed on well managed pasture.

At the University of Vermont, for example, researchers recently measured low levels of undegradable protein (8.8 to 21.3% of total protein) in forage of a well managed pasture throughout the grazing season. They fed two concentrate rations to Holstein and Jersey milking cows that were getting all of their forage from the pasture. The rations were: 1) a high bypass protein formula containing 61 percent of total concentrate intake protein as undegradable protein; and 2) a low formula containing 23 percent of total protein as undegradable protein, plus energy in the form of 6.5 pounds of high moisture ear corn. Milk production increased 11 percent with both concentrates. Stated another way, the increase in milk production from high undegradable intake protein concentrate equaled the increase in milk production from 6.5 pounds of high moisture ear corn. These results could be important, because if an inexpensive ration adjustment in undegradable protein increases milk production from cows on pasture, a major economic benefit is possible for dairy farmers using pastures. Research is continuing to determine more precisely how much undegradable intake protein needs to be in concentrates for supplementing cows on pasture (J. Welch, personal communication).

Energy

Since protein deficiencies usually don't exist for most animals (except for high-producing dairy cows) grazing well-managed pastures, such pastures may be evaluated only in terms of the amount of energy in the feed that's available to the animals after digestion and metabolism.

In terms of quantity, energy is the most important part of an animal's diet. All feeding standards and rations are based on some measure of energy, with additional inputs of protein, amino acids, essential fatty acids, vitamins, and minerals.

When there's plenty of forage to graze (adequate forage allowance), the amount of energy in the forage that's available to animals is related to the animals' intake of the forage. Intake is greatest for pasture forage that's low in fiber and high in protein and energy. Such forage is very palatable to grazing animals. As pasture forage quality decreases, the bulk of the forage eaten limits intake, because the larger amount of fiber in the plants breaks down more slowly and keeps the gastrointestinal tract full of fiber longer. (See Chapter 8 for further discussion about energy in livestock production.)

MILKING COWS

More than anything else, feed determines the profitability and productivity of dairy cows. Feeding expenses account for 45 to 65 percent of the total cost of producing milk. The lower these costs are, the more profitable the dairy farm will be. Pasturing is the lowest cost way of feeding dairy animals and other livestock.

Depending on pasture forage energy content and liveweight of the cow, certain amounts of milk and butterfat can be produced without feeding any concentrates. Cows producing the most milk require excellent quality pasture forage, to achieve the needed intake of energy for high production levels without supplements (see Table 8-9). Smaller cows can't achieve as high a level of production on poor forage as larger cows can. This probably is due to the physical limitation of how much forage can be eaten in a day by different size animals. But smaller cows produce more milk per 1,000 pounds of body weight than larger cows, and are therefore more efficient than larger cows. Smaller cows (e.g. Jerseys) also make up for producing a lower volume of milk than larger cows (e.g. Holsteins) by putting more butterfat in their milk and eating less total feed.

Another important point to note here for milking cows and all grazing animals, is that at a particular level of productivity, a greater intake of energy is needed when animals graze low-quality pasture, compared to that needed when they graze high-quality pasture. This is because they walk around less searching for preferred feed, and expend less energy both in obtaining the feed and in digesting it, when they graze high-quality pasture.

Along with energy, dry matter intake is extremely important in feeding dairy cows. Cows will eat about 3 to 3.5% of their bodyweight in dry matter per day. For cows ranging in weight from 800 to 1300 pounds, that amounts to 24 to 46 pounds of dry matter. Since pasture forage contains about 75 percent moisture, a cow must eat 100 to 200 pounds of high-quality, palatable green pasture forage per day! Grazing management must make it as easy as possible for the cow to eat that much.

The more milk a cow produces, the more digestible fiber she needs in the total ration to sustain a maximum percentage of butterfat in the milk. But when the proportion of fiber in the ration rises above a certain level, milk production drops. For each cow, depending on the size of its rumen, a point exists where milk production can only increase by decreasing the percentage of butterfat in the milk, and butterfat can be increased only by decreasing milk production.

Fiber in forage stimulates the rumen and increases formation of acetic acid, a precursor of milk fat. Fiber also promotes cud-chewing and secretion of saliva, which acts as a buffer in the rumen and favors production of acetic acid. Fiber may also slow passage of high-quality pasture forage material through the rumen, resulting in more complete digestion and use of its protein and energy.

During many years of selecting cows in the United States to use large amounts of concentrates, they may have been bred away from efficient use of pasture forage. Because of this, you may have to feed 1 to 2 pounds of hay or other equivalent roughage at each milking throughout the grazing season to minimize depression of milk butterfat content from some cows. For example, one farmer in our 1984 study found that by feeding 3 to 13 pounds (depending on level of milk production: see details in Chapter 10) of coarsely ground, high-moisture ear corn at each milking instead of hay, milk butterfat contents remained high and all energy needs were met at the same time.

The pattern of dry matter intake by cows over the season probably will differ somewhat from those shown in the feed planning examples given in Chapter 8. This is because it is nearly impossible for milking animals to eat and process the amount of nutrients required to meet the needs of the udder in producing large quantities of milk. Milking animals mobilize their body fat reserves in early lactation to satisfy energy needs when demand is high. In mid and late lactation these reserves are replaced by eating forage in excess of their needs, and dry matter intake becomes balanced over time, as long as allowance of high-quality forage is adequate. If cows don't have enough high-quality forage available, they lower their needs by reducing milk production, unless they are fed protein and energy supplements.

Cows should be dried off 45 to 60 days before freshening. During the dry period, cows can be grazed on pasture containing less energy (30 percent over maintenance needs) than that shown in tables in Chapter 8. In fact, grazing a forage containing less energy may help to dry cows off and prevent them from becoming too fat. A good way to lower the energy content of the forage available to dry cows, is to have them graze paddocks after milking cows have eaten the highest quality forage. (Jersey dry cows may need to be removed from pasture a month before calving, to avoid milk fever. Even when dry cows follow milking cows to clean up paddocks, the forage may still contain too much calcium and too little fiber for Jersey dry cows.)

Replacement heifers can live on pasture alone after they are about 6 months old. They also can be grazed on forage containing somewhat less energy than that needed by milking cows. Feeding too much energy to heifers before breeding actually results in them producing less milk. Heifers should be fed to breed at 13 to 14 months old, so they calve when they are 22 to 24 months old. Thirty to 45 days before calving, heifers should begin grazing the same high-energy forage as milking cows receive. Until then, heifers can follow milking cows through paddocks to eat remaining forage that contains less energy.

BEEF CATTLE

One of the most disgusting things that I have ever seen and smelled was a beef cattle feedlot off of Interstate 80 in Nebraska. I'm sure that if people knew their beef was coming out of cesspools such as that, less beef would be eaten. Feedlots now produce more organic waste than the total amount of sewage from all municipalities in the United States! Feedlots operate at a profit only because they are allowed to pollute the environment and pass those costs on to society, which sooner or later has to pay them in some way. It's another example of economics as if people, animals, quality of life, and the Earth don't matter.

Feedlots really can't even be justified on economic terms. Feed certainly affects total productivity and profit in beef operations, accounting for about 75 percent of the cost of keeping beef cows, and raising and finishing beef cattle. But grazing them on pasture can cost about six times less to accomplish those objectives than it does in confinement! And that doesn't include the savings of about 2 bushels of topsoil that erode away in producing each bushel of corn, which is used to "finish" animals to the point that their meat is too fat and causes health problems in people that eat it. I wonder how much money is spent each year in this country just trimming fat from feedlot-finished carcasses.

It is true that the faster beef animals grow the more efficient the production is, in terms of forage dry matter, protein, and energy needed, because the daily effect of the maintenance requirement on total production is reduced. For example, 1,200 days, 16,140 pounds of dry matter, and 1,438 pounds of crude protein are needed for a steer gaining 0.55 pounds per day to grow from a 350-pound starting weight to a 1,000-pound market weight. In contrast, a steer gaining 1.65 pounds per day reaches the 1,000-pound market weight from a 350-pound starting weight in 400 days, and only needs 6,730 pounds of dry matter and 684 pounds of crude protein to do it! But high rates of daily weight gain are possible to achieve in other ways than confining animals in feedlots. Feed-cost savings can be very large when beef cattle graze well managed pasture, because daily weight gains of 2.75 to 3.30 pounds are possible during at least 6 months of the year.

Feeding beef cattle as much high-quality forage as they can eat also results in large savings of pasture forage if rapid growth is maintained. This allows more animals to be carried on a pasture, and increases the profitability of the operation.

In addition to information in tables in Chapter 8, note that dry pregnant mature beef cows need about the same amounts of energy as those given for dry dairy cows. Beef cows nursing calves need 3 to 5 Mcal more energy per day than they needed before calving.

SHEEP

Mature sheep spend almost their entire lives at maintenance level, which means that their energy needs are more or less constant, except during certain periods in gestation and lactation. Liveweights of ewes fluctuate during periods when feed quality is adjusted to the animals' needs. The energy requirement of adult ewes during the last 4 weeks of gestation is 1.5 times their maintenance need. This increases to 3 times maintenance during early lactation, and decreases to 2 times maintenance by the third month of lactation. After weaning, ewes can return to maintenance, until flushing.

During maintenance periods ewes can tolerate forage with lower energy contents, so they can be used during such times to clean up paddocks after lambs or other livestock, or in mob-stocking of rough land for pasture improvement (see Chapter 11). Be careful in using mob-stocking so that ewes aren't kept on poor-quality feed for too long, though, or a reduced number of lambs may be born the next spring. It's best to alternate a day of grazing low-quality pasture with a day of grazing higher-quality pasture, to avoid stressing the animals too much. Grazing ewes on forage that is better than their minimal need results in them weighing more and consistently giving birth to more and larger lambs that gain weight faster.

Lambs

For a certain rate of liveweight gain, lambs need a greater energy intake if they are grazing low-quality pasture, compared to that needed when grazing high-quality pasture. Fast rates of gain can't be achieved with low-quality pasture, because the bulk of feed in the rumen limits intake before enough energy has been ingested to reach high weight gains. Therefore, the energy of low-quality pasture is used less efficiently than that of high-quality pasture. This difference between low- and high-quality pasture becomes more obvious as daily weight gain increases, in terms of energy requirements for good performance.

Lambing in spring enables ewes to be fed on high-quality forage during most of their lactation, resulting in considerable savings from not having to feed concentrates to produce the milk. Lambs begin to graze soon after being placed on pasture with their mothers, and the forage they eat stimulates rumen development, so they progressively make more use of forage.

Lambs have the fastest rate of gain on pasture while nursing their mothers' milk. Pregrazing pasture mass should be 2000 to 2200 lb DM/acre (about 4 inches tall) for ewes with lambs or lambs alone. Postgrazing pasture mass shouldn't get below 1,200 lb DM/acre (1.5-2 inches tall) for best milk production and lamb weight gain.

Ideally, lambs reach market weight at 14 to 16 weeks of age from mother's milk and pasture forage, without any concentrate feeding, and can be taken directly to market at that time. Lambs that haven't reached market weight or are being kept as replacements can be weaned and separated from their mothers at 14 to 16 weeks of age. At this time, even if they aren't separated from the ewes, lambs should be treated to kill parasites that they picked up while grazing with their mothers. Ewes should be wormed at lambing, and may be wormed again at this time and in the fall, although New Zealand farmers have found no advantage to worming adult ewes more than once a year right after lambing.

It's best if weaned and wormed lambs begin grazing on pasture that was set aside in May and machine-harvested in June. In this way lambs enter a pasture area that wasn't grazed and didn't receive parasite eggs during the current season and, consequently, the lambs may have less parasites for most of the season. After the first worming, lambs should be wormed twice at no more than 21-day intervals, and then every 28 days until the end of the grazing season or until they are marketed.

Other possible ways of controlling worms in ewes and lambs on pasture besides individually treating animals with anthelmintics are: 1) feeding diatamaceaous earth mixed half and half with salt; 2) providing drinking water that contains 30 parts per million of food grade hydrogen peroxide; or 3) alternate areas grazed by sheep with those grazed by cattle, either within a season or preferably in different years. Carefully monitor your flock when using these methods without anthelmintics to control worms, to be certain that animal parasite levels actually are low.

All of the above practices ideally promote high, sustained daily weight gains of lambs, while minimizing production costs and maximizing profits. Unfortunately, wholesale lamb markets in the United States require a finish and size of lamb that may be difficult to obtain on pasture forage alone, given the sheep that commonly exist here. Achieving 16-week-old lamb liveweights of 85 to 90 pounds on pasture without grain supplement is relatively easy to do. But it's more difficult to consistently get lambs up to the 105- or 110-pound liveweight required by current markets, without grain feeding .

Whole corn or some other energy supplement can be fed to lambs on pasture, beginning just before the end of the grazing season, but it doesn't make any sense. Judging from French market requirements, the ideal lamb slaughter liveweight range is 80 to 88 pounds. This is because at that weight the meat is more lean, tender, and tasty. As lambs are raised to 105 pounds on grain, they do get bigger, but they also get fatter and less tender. Who wants to eat fat lamb? This requirement for bigness may disappear as producers and consumers realize the benefits of lean lamb grown entirely on pasture without grain.

Obviously, this realization will have to be preceded by more widespread development and use of suitable high-quality pasture, along with appropriate selection and breeding of sheep. The big long-legged sheep currently used in the USA compare poorly with New Zealand, Irish, Scottish, and British lines selected for efficient growth on pasture.(D. Flack, personal communication)

GOATS

Goats can adapt to a wide range of conditions, but they have certain unique nutritional characteristics, which require that they be treated and fed differently from cows or sheep for best results. Despite the jokes about goats eating tin cans, their nutritional needs actually are higher than those of other ruminants. Milking goats will eat up to about 5.5 percent of their body weight in feed each day. In contrast, cattle and sheep usually eat 3 to 3.5 percent of their body weight in feed per day.

Probably because of their high nutritional needs, goats are very selective grazers, and will use a pasture poorly unless managed well. Goats graze pasture forage from the top downwards in layers. Their level of dry matter intake and performance is very sensitive to pasture height and mass. Compared to young sheep, dry matter intake of young goats decreases at a faster rate as pasture mass declines. Goats appear to stop grazing at a pasture mass of about 900 lb DM/acre (about 1 inch tall), so don't attempt to have them graze lower than that. Research results are very limited, but one study showed that a forage allowance of 15 lb DM/head/day resulted in maximum productivity. Always feed or have some hay available to goats grazing lush well-managed pasture, to satisfy their need for browse.

Does must have a dry period of at least 50 days before kidding. As soon as does are dried off, they can be maintained on pasture forage containing less energy than they were eating while milking. Feeding woody browse at this time is also helpful in restoring the function and capacity to their rumens and digestive systems, which may have changed a great deal and decreased in size during the months when they grazed only high-quality forage. During this time, does can be used to graze rough, brushy pastureland that needs improving. Be careful that does aren't stressed when mob-stocking rough land.

Kids should be allowed to nurse their mothers for at least 4 days before being weaned to a bottle. This helps to condition the mothers' udders and gives the kids the best start possible. Make certain that does evenly milk out. (A virus transmitted through milk currently is a major health problem in goats. If it's present in the herd, all colostrum and milk should be heat-treated before feeding to kids.)

In dairy herds, kids can be given milk replacers at 4 to 7 days of age so that the mother's milk can be sold. When separated from their mothers for weaning, keep kids in small pens where they can be seen, smelled, and heard, but not nursed by the mothers for about 10 days. After this time they can run with their mothers without any

185

problem of nursing the mothers. If you don't do something like this, you'll be driven crazy by the mothers who will devise any means to get you to substitute for their kids. Even if you can ignore the noise and commotion of the mothers, all of their activities require energy, and that energy could be put to better use in producing milk.

Kids should be offered forage when they are about 2 weeks old or less, to help develop their rumens and make weaning easier. Between 2 weeks of age and weaning, a kid needs 1 to 2 quarts of milk or milk substitute per day, plus pasture forage, water, salt, trace minerals, and sunshine. Kids can be weaned at 6 to 8 weeks or as soon as they reach 2.5 times their birth weight. The earlier the weaning before they reach this size, the more their growth rate slows. To have no slowing of growth rate, weaning should be very gradual, and not begin until kids are at least 6 weeks old.

In herds kept for meat production, female kids needed for reproduction should be grazed with their mothers during as much of the milk feeding period as possible and not weaned early. In dairy herds, female kids grazing well-managed pastures reach liveweights of 70 pounds by the time they are about 7 months old; then they can be bred at about 8 months of age. Kids raised for slaughter should be weaned when they weigh 45 to 55 pounds, depending on the market. (A. Pell, personal communication)

HORSES

Of all livestock, horses especially do well when grazed on good pasture. The freedom and exercise they experience under pasture conditions develop strong bones, muscle, and spirit.

The efficiency and years of service of horses are largely

determined by how well their nutritional needs are met. Like other animals, horses require nutrients for maintenance, growth, reproduction, and production, which in horses is for working. Unlike other animals, the work of horses generally is very strenuous and irregular, which stresses them and makes it difficult to feed them according to their needs. Another complication is that the nutritional needs of horses may not remain the same even from day to day. Besides a horse's health, condition, and temperament, many factors influence the nutritional needs of horses. These include: age and size, stage of gestation or lactation, kind and amount of activity, weather, and quality and quantity of feed available. So besides feeding tables used as guidelines, a lot of experience, skill, and good judgment are needed to feed horses properly.

Well managed pastures, with salt and minerals available free-choice, can provide most or all of the nutrients needed by horses. A mineral mixture containing two parts calcium to one part phosphorus should be available if the pasture is mainly grass. If the pasture contains mostly legumes, the mineral mixture should be no more than one part calcium to one part phosphorus.

PIGS

High-quality pasture can satisfy much of the nutritional needs of pigs at all stages of gestation, lactation, growing, and finishing. Since 75 to 95 percent of pig feeds used in the United States are derived from grains, large savings in feed costs can be realized by grazing pigs on well managed pastures. Often pigs can provide greater dollar returns from an acre of pasture than any other kind of livestock.

Pigs on pasture develop a surprisingly large capacity for using forage when the feeding of supplements is limited, so they're forced to graze. Unlike ruminants, however, the monogastric pig must rely mainly on feeds having readily digestible carbohydrates to meet energy needs. The more complex carbohydrates (e.g. cellulose and hemicellulose) contained in forages are broken down only by microbial fermentation. Because the pig doesn't have a rumen where such fermentation could occur, fibrous components of forage aren't used efficiently by pigs. For this reason, pigs usually can't be grown or fattened satisfactorily on pasture forage alone.

But pasturing of pigs has several real advantages over drylot feeding:

1. Feed savings. Well managed pastures can save large amounts of grain and protein supplements. With properly balanced rations the feed savings have been found to be: a) 10 to 20 percent less grain and 30 to 50 percent less protein supplements needed for producing 100 pounds of pork; b) 500 to 1000 pounds of grain and 300 to 500 pounds of protein supplement saved per acre of pasture; c) 50 percent lower feed costs for brood sows; and d) about 3 pounds of mineral supplement per acre of pasture.

2. Pastures decrease nutritional deficiencies in pigs, because of the higher quality proteins, vitamin content, unknown growth factors, minerals, and antibiotics that pastures provide in comparison to drylot feeds.

3. Pasturing pigs reduces labor costs by about 30 percent. This results from less effort and expense in sanitation, especially in manure handling, and less labor needed because the pigs harvest part of their own feed. No harvesting method is as efficient as allowing animals to graze their feed.

4. Pastures conserve the maximum nutrient value of manure. At least 80 percent of nutrients eaten are returned directly to pasture soils.

5. Grazing sows on well managed pasture during their

pregestation period is especially important. The sows don't become overly fat and have greater reproductive efficiency. Pastured sows have 1.5 times more pigs born per litter than sows fed a similar ration in drylot.

6. Because pasture forages contain large amounts of protein, energy, vitamins, and minerals, they provide excellent feed for sows during gestation. This is especially true if the forage mixture contains 7 to 10 percent of deep-rooted plants, such as chicory, dandelion, and yarrow, which have high mineral contents. Besides these nutrients, many forages contain reproductive factors needed for good litter size and survival of newborn pigs.

7. Pastured pigs grow more vigorously and there are fewer runts. This is due to a combination of the green succulent feed, proper nutritional balance, exercise, and better sanitary conditions with less incidence of diseases and parasites.

8. Well managed pastures also provide excellent feed for sows and promote milk flow for their litters. Pigs nursed on pasture have a higher survival rate and are larger at weaning than pigs nursed in drylot.

9. Weaned pigs on excellent pasture gain weight faster and reach market weight 1 to 2 weeks earlier than those fed in drylot. Contrary to opinion, the restricted exercise of animals fed in confinement doesn't result in more rapid and efficient liveweight gains.

What all of this means is that pigs can be produced on well managed pasture, with large feed savings and, consequently, greater profitability than in drylot.

Pigs have a natural tendency to dig and root in soil. If they have plenty of good-quality forage to eat, they tend to root less. But to use Voisin controlled grazing management properly, pasture forage must be grazed down to 1 to 2 inches from the soil surface in each occupation period. Forcing pigs to graze so close to the ground may cause them to root up the soil. If they begin to dig, the pasture must be protected by putting rings in their noses, to keep them from rooting up the plants and generally destroying the pasture plant community.

Parasite levels must be kept low in pastures used by pigs. This can be accomplished if every 2 or 3 years a pasture is rested completely from grazing by pigs for 2 years, to allow parasites time to die in the absence of their host. During that time the pasture can be grazed by other livestock or machine-harvested.

POULTRY

Poultry produced on pasture? Of course! That's the way it is still done on small, diversified farms, and that's the way it used to be done on all farms before the trend toward confinement feeding of all livestock began. Confinement feeding of livestock was encouraged by the availability of inexpensive grains. Economically I suppose that made sense, but from the viewpoints of ecology, soil erosion, food quality, and responsibility for the world community, it never makes any sense to feed grain to livestock unnecessarily. Grain never really is cheap to produce in terms of energy used and costs of soil nutrients and erosion. Isn't it a shame that people are led around by the nose by economics (that doesn't take into account social and environmental costs), rather than doing things in ways that just feel right?

Chickens, turkeys, ducks, and geese can use pasture forage to the same advantage as other livestock, as long as the pasture and poultry are well managed. In many cases, the return per acre of good pasture used by poultry will be more than from its use by other livestock. The number of birds able to be carried per acre of pasture varies, depending on kind, quality, and quantity of forage available, weather, and soil texture and fertility.

Careful attention must be paid to grazing height, because chickens and other poultry can easily graze plants too low. Their grazing should be managed so that they don't graze lower than 3/4 to 1 inch from the soil surface. Occupation periods must be short to prevent overgrazing and digging of the pasture.

Provide shade and shelter on the pasture so that poultry can get out of direct sunlight and rainfall when they want to, and have a place to lay eggs in and roost at night, protected from predators (e.g. cats, dogs, foxes, skunks, weasels, coyotes, raccoons).

Movable coops on wheels can fill all of the above needs. A rolling henhouse following cattle through their pasture rotation (see Joel Salatin in References) is a good example of how poultry can be pastured. A 6- x 8-foot coop on bicycle wheels can be used for small flocks, either confined to paddocks or allowed to range freely. Hexagon-shaped paddocks can be formed by two pairs of three 10- x 4-foot panels made from 3/8-inch iron rod and chicken wire. By rotating the paddock around the mesh-floored coop, several days' of fresh pasture can be gotten per move of the coop. For larger flocks (e.g. 100 layers) a 12- x 20-foot coop on wheels can be moved with a tractor to pasture the free-ranging flock. The hens range up to about 200 yards away from the coop. Besides dramatic savings in feed costs, poultry grazing behind cattle can benefit overall pasture production. By eating insects, especially developing fly larvae in cattle manure, poultry help reduce fly and other insect problems. By breaking down manure pies in their search for fly larvae, poultry enable nutrients contained in the manure to cycle faster, and zones of repugnance to disappear sooner, resulting in less rejected forage and more net forage production.

CHAPTER 7

Chickens

Most people don't even know what good chicken or eggs taste like, because they haven't eaten any that were produced on pasture. I know that the things produced in confinement are called chickens and eggs, but that's where the similarity ends. You can do it another way.

Actually, there are very good economic reasons for producing chickens and eggs on pasture:

1. It may cost only half as much to feed chickens for growth and egg production on well managed pasture, as it does in confinement.

2. Money also can be saved in housing and management. By keeping hens on pasture with portable coops for laying and roosting until they finish laying in the fall, laying quarters in the main coop(s) can be made available for a new crop of spring pullets.

3. When pullets are 7 to 10 weeks old, they also can be grazed on pasture, with large savings of feed possible. Pullets on pasture gain weight more rapidly, need less feed per pound of gain, and have better vigor than pullets raised in confinement.

Important considerations of well managed poultry pasture, besides the plants available, are that pasture soils be free of poultry disease contamination, and that young birds graze well isolated from older birds to prevent spreading diseases. Rotating birds through the pasture and periodic (every 2 years at least) resting of pasture areas completely from poultry grazing (machine-harvest the areas or graze with other livestock) for 2 or 3 years, should prevent soils from becoming contaminated with diseases.

All legumes (especially white clover), grasses, and edible "weeds" (e.g. plantain, dandelion, lambsquarters) commonly present in pastures provide excellent feed for chickens and other poultry. All required protein, vitamins, energy, and most minerals can be provided by well managed pasture. Plants alone may not provide all needed amino acids, but grazing poultry also eat insects and worms, which supplement the plant diet. The only minerals that must be provided are sodium and chlorine, and possibly calcium and phosphorus. These can be provided easily by making available free-choice in separate containers common loose salt (sodium chloride), oyster shell or ground limestone, and defluorinated rock phosphate. Water, of course, must always be available.

Chickens on pasture must be managed so that they use forage to the fullest extent possible. (For earliest maturity and highest initial level of egg production, feed for pullets on pasture should only be restricted by about 15 percent of what they would receive in confinement.) Any feed that's provided should be considered as a supplement to pasture forage; otherwise chickens will depend on the feed rather than forage. Supplements must be appropriate for the kind and quality of forage available. Chickens can be grown and eggs can be produced quite successfully on pasture with only supplements of minerals and small amounts of whole wheat or corn.

Turkeys

All of the above applies equally to producing turkeys on pasture, except that turkeys must be completely isolated from chickens. This is mainly to prevent infection of the turkeys with black-head disease, which can be transmitted by infected carrier chickens. If chickens have grazed a pasture, allow a 2-year interval before using it for turkeys.

Ducks and Geese

Everything stated for producing chickens on pasture also

applies to ducks and geese, except they don't need any shelter from the rain; instead they need a plentiful source of clean water for bathing. Of all poultry, geese and Muscovy ducks are the best foragers and easiest to keep on pasture.

REFERENCES

Belanger, J. 1975. *Raising Milk Goats the ModernWay*. Garden Way Pub., Charlotte, Vermont.152 p.

Berry, W. 1981. *The Gift of Good Land*. North Point Press, San Francisco. 281 p.

Blowey, R.W. 1985. *A Veterinary Book for Dairy Farmers*. Farming Press, Suffolk, England. 397 p.

Church, D.C. 1984. *Livestock Feeds and Feeding*. O & B Books, Corvallis, Oregon. 549 p.

Commoner, B. 1972. *The Closing Circle*. Alfred A. Knopf, New York. 326 p.

Cromwell, G. 1984. Feeding swine. p. 389-412. In: D.C. Church (ed.) *Livestock Feeds and Feeding*. O & B Books, Corvallis, Oregon.

Ensminger, M.E. 1952. *Swine Husbandry*. Interstate Printers and Publishers, Danville, Illinois. 378 p.

Ensminger, M.E., and C.G. Olentine, Jr. 1978. *Feeds and Nutrition*. Ensminger Pub., Clovis, California. 1417 p.

Flanagan, J.P., and J.P Hanrahan. 1987. *Lowland Sheep Production*. An Foras Taluntais, Belclare, Tuam, Ireland.

Guss, S.B. 1977. *Management and Diseases of Dairy Goats.* Dairy Goat Pub., Scottsdale, Arizona. 222 p.

Henderson, D.C., H.C. Whelden, Jr., and G.M. Wood. 1953. *Range and Confinement Rearing of Poultry.* University of Vermont Agricultural Experiment Station Bulletin 568.

Hughes, H.D. and M.E. Heath. 1967. Hays and pastures for horses. p. 671-683. In: H.D. Hughes, M.E. Heath, and D.S. Metcalfe (eds) *Forages.* Iowa State Univ. Press, Ames.

Jackson, W. 1980. *New Roots for Agriculture.* Friends of the Earth, San Francisco. 155 p.

Jagusch, K.T. 1973. Livestock production from pasture. pp. 229-242. In: R.H.M. Langer (ed.) *Pastures and Pasture Plants.* A.H. and A.W. Reed, Wellington, New Zealand.

Kruesi, W.K. 1985. *The Sheep Raiser's Manual.* Williamson Pub., Charlotte, Vermont. 288 p.

Kennard, D.C. 1967. Forage for poultry. p. 663-670. In: H.D. Hughes, M.E. Heath, and D.S. Metcalfe (eds) *Forages.* Iowa State University Press, Ames.

Mackenzie, D. 1957. *Goat Husbandry.* Faber and Faber Ltd., London. 368 p.

McCall, D.G., and M.G. Lambert. 1987. Pasture feeding of goats. p. 105-109. In A.M. Nicol (ed..) *Feeding Livestock on Pasture.* New Zealand Soc. Animal Prod., Hamilton.

Morand-Fehr, P. and D. Sauvant. 1984. Feeding goats. p. 372-388. In: D.C. Church (ed.) *Livestock Feeds and Feeding.* O & B Books, Corvallis, Oregon.

Mott, G.O., and C.E. Barnhart. 1967. Forage utilization by swine. p. 655-662. In: H.D. Hughes, M.E. Heath, and D.S. Metcalfe (eds) *Forages.* Iowa State University Press, Ames.

Nicol, A.M., D.P. Poppi, M.R. Alam, and H.A. Collins. 1987. Dietary differences between goats and sheep. Proceedings of New Zealand Grassland Association. 48:199-205.

Poincelot, R.P. 1986. *Toward a More Sustainable Agriculture.* AVI Publishing Co. 241 p.

Poppi, D.P., T.P. Hughes, and P.J.L'Huillier. 1987. Intake of pasture by grazing ruminants. p. 55-63. In A.M. Nicol (ed..) *Feeding Livestock on Pasture.* New Zealand Society of Animal Production, Hamilton.

Sampson, R.N. 1981. *Farmland or Wasteland: a Time to Choose.* Rodale Press, Emmaus, Pennsylvania. 422 p.

Simmons, P. 1976. *Raising Sheep the Modern Way.* Storey Communications, Pownal, Vermont. 234 p.

Subcommittee on Beef Cattle Nutrition. 1984. *Nutrient Requirements of Beef Cattle.* National Academy of Sciences, Washington, D.C. 6th ed.. 55 p.

Van Soest, P.J. 1982. *Nutritional Ecology of the Ruminant.* O & B Books, Corvallis, Oregon. 374 p.

Voisin, A. 1959. *Grass Productivity.* Island Press, Washington, D.C. 353 p.

8

Feed Planning

A cold rain starting
And me without a hat.
On second thought, who cares?
                    Basho

Achieving any goal requires planning. Feed planning helps increase farm profitability through more effective use of feed resources. In planning the feeding of grazing livestock, feed requirements are compared with feed supply from pasture and supplements. You can make feed plans that vary from long term of a year or more, to short term of 1 day. Three kinds of feed plans are used for specific decisions and time period:

1. Feed profile: a long term plan of a grazing season or longer that is used for making decisions such as stocking rate, calving or lambing date, weaning time, synchronized drying off date, and probable need of stored feed or supplements. Feed profiles are based on the average or most likely pattern of feed demand and supply.

2. Feed budget: a medium term plan of up to 6 months long, used to balance feed demand with feed supply for an

existing situation. Feed budgets frequently involve decisions about the best or most profitable use of pasture surplus forage, or the least expensive way of making up for insufficient (deficit) pasture forage.

3. Grazing plan: a short term plan that involves deciding how long groups (mobs) of livestock should graze paddocks. The grazing plan helps to determine how long the rotation will be.

Feed plans involve estimates of feed demand and supply. As with all plans, you must monitor, control, and replan if necessary, to achieve your goal. With feed plans you must continually monitor pasture and livestock production, and make any adjustments that are needed because of changes from your predictions. Planning pasture feeding of livestock may appear to be imprecise and sometimes inaccurate because of wide variations from estimates of pasture plant growth rate, pasture cover, forage feeding value, or livestock production levels. But feed planning can and should be the basis for organizing pasture feeding of livestock.

FEED SUPPLY

Pasture plant growth rate, pasture cover, and supplements make up the feed supply. The contribution of each part differs depending on the kind of planning that you do.

Plant Growth Rate And Total Forage Production

To plan ahead, you need to estimate the growth rate of pasture plants during the period of your feed plan. Unfortunately, very little information exists about average growth rates of pasture plants under controlled grazing management at different locations in the United States. Data for locations in Vermont and West Virginia in Table 8-1 and Figure 8-1 are examples of needed information. Average regional growth rates are a good starting point for your planning. But you need to estimate them for your farm and even for particular pastures, because plant growth rate depends on climate, soil fertility and moisture, slope aspect, and plant species.

Grazing management also influences plant growth rate and overall pasture production. For example, low stocking density with little or no machine harvesting of surplus forage (lax grazing management) usually results in high pasture masses and less net forage production, due to increased losses from death and decay. In contrast, severe grazing may reduce new plant growth and, consequently, decrease both total and net forage production.

Plant growth rate can be roughly estimated from changes in pasture cover and estimated net forage production. These data collected over several years can provide a good indication of the likely seasonal range of forage production in individual paddocks and the whole pasture. Plant growth rate is the most variable part of medium- and long-term feed plans, and therefore your plans may need to change as seasons progress. Ignore plant growth rate for 1- to 7-day feed plans.

Table 8-1. Average plant growth rates in grass-white clover permanent pastures under Voisin controlled grazing management (1989-1990)

Month	Vermont	West Virginia
	----lb DM/acre/day----	
April	33	53
May	59	64
June	57	70
July	42	54
August	40	36
September	33	26
October	12	18

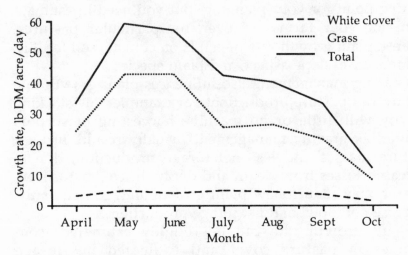

Figure 8-1. Seasonal pattern of grass, white clover, and total net forage production in pasture under Voisin controlled grazing management near Colchester, Vermont during 1989-1990.

Table 8-2. Example of average pasture cover calculations.

Paddock number	Pasture mass per paddock	Paddock area	Total forage
	lb DM/acre	acre	lb DM
1	1650	1.0	1650
2	2000	1.5	3000
3	2200	1.6	3520
4	1250	1.9	2375
5	1850	1.0	1850
6	2550	1.0	2550
	Total	8.0	14945

Average pasture cover = 14945/8.0 = 1868 lb DM/acre.

Pasture Cover

Pasture cover is an estimate of average total pasture forage on a farm at a certain time. Calculate it by multiplying pasture mass in each paddock by the paddock area to estimate total forage in the paddock. Add the amounts of forage in individual paddocks together, then divide by total pasture area, to get average cover (Table 8-2 above).

As discussed in Chapter 5, pasture mass can be estimated by clipping samples to soil surface level, and drying and weighing the forage samples. Since this is very time consuming, pasture mass and cover usually are estimated visually or by using a bulk height plate, pasture probe, or ruler. Of course, these indirect measurements require calibration with clipped forage samples. With experience you can visually estimate pasture mass to within 100 to 300 lb DM/acre of the actual amount. Forage in short swards is more accurately estimated visually than that in long swards over 6 inches tall, which usually is underestimated. But any errors in estimating pasture mass usually are small compared to those in predicting plant growth rate or knowing actual paddock areas.

Since pasture mass is estimated to ground level, so is pasture cover. If there is more than 25 percent dead material in the sward, pasture mass should be estimated as pounds of green dry matter per acre, because animals won't eat the dead material.

Suitable pasture cover on any farm depends on such things as expected pasture plant growth rate, stocking rate, and timing of lambing or calving. Usually you should try to maintain average pasture cover above 900 lb DM/acre, and less than 2200 lb DM/acre. The pasture cover of your feed plan determines if forage allowance or postgrazing pasture mass associated with planned forage dry matter intake can be achieved.

Be sure to monitor pasture cover regularly, since it

indicates if your feed plan is OK or needs changing. A good way to keep track of the whole schmeer is to graph the average pasture cover expected at the end of each month (Figure 8-2). then you can add the actual cover values that you estimate as the season progresses. If the actual value is much lower than the expected one, there might be a feed deficit that may not be obvious yet. This method can provide a warning of a future problem that is developing, before it becomes serious, and enable you to take action to avoid the problem.

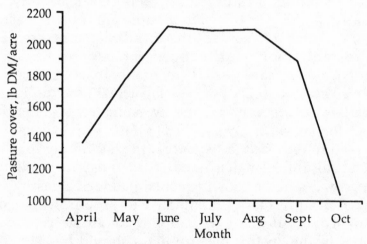

Figure 8-2. Example of expected pasture cover (values from Table 8-2).

Supplements

Supplemental feeds such as silage, green chop, hay, or grain concentrates form part of the total feed supply. Supplements substitute for pasture, depending on their protein and energy feeding values, so be careful in feeding them because they can reduce pasture forage intake.

Harvesting and storing excess forage as hay or silage reduces pasture surpluses that occur when plant growth rates are greater than feed needs. Machine-harvesting surplus forage decreases pasture cover in areas that are harvested, and helps maintain the overall pasture in good condition for subsequent grazings. Storing excess forage is a way of carrying forward surpluses to times of shortage. Stored surplus forage can be used to reduce pasture feed deficits during the grazing season, fed out in winter, or sold to others to meet their feed deficits.

FEED DEMAND

Forage Quality: Protein And Energy

In estimating feed demand, New Zealand farmers disregard pasture protein content, because they know from experience that it's never deficient for their animals grazing their well-managed pastures, so there's no point in continuing to be concerned about it. Instead, they evaluate pasture in terms of metabolizable energy (ME), which is the gross energy of a feed as it is eaten, minus the energy lost in manure, urine, and gases.

For most practical situations, New Zealand farmers classify pasture forage in terms of Megacalories (Mcal) of ME per kg of DM as: poor (1.8), fair (2.2), good (2.6), and excellent (3.0). Of course forage usually won't contain exactly 3.0, 2.6, 2.2, or 1.8 Mcal ME/kg DM. These are midpoints of forage ME-content ranges. Specifically, forage containing more than 2.8 ME is excellent, 2.4 to 2.8 is good, 2.0 to 2.4 is fair, and less than 2.0 is poor. This method permits easy use of forage energy values for feed planning under normal, changing feeding conditions.

If everything was equal and people agreed about things in general, this same idea could be used without any problem in North America and elsewhere. But all things aren't equal and people don't agree about much of anything. We especially in the United States use a discombobulated system of weights and measures, that's neither this nor that. But this doesn't mean we can't use the New Zealand method of evaluating pastures to our advantage. Let's see how we might do that.

Energy value of forages in the United States is expressed in various ways, including percent total digestible nutrients (TDN), or Mcal per pound (!) of DM digestible energy (DE), net energy maintenance (NEM), net energy gain (NEG), net energy lactation (NEL), or metabolizable energy (ME). Most net energy values have been calculated from older data from feeding trials in which energy was originally expressed as TDN, DE, or ME. Very few feeds have been evaluated directly in terms of NEM, NEG, or NEL. Although the use of net energy values currently is popular in the United States, it should be noted that recalculating new values from existing data doesn't necessarily make the new complicated values better than the old simple ones. (Because of this, there is a rumor circulating that some American animal scientists favor returning to using simple ME and TDN values.)

When I asked an eminent animal scientist which energy value would be best to use in the United States for planning the feeding of livestock on pasture, he said, "Oh, you don't want to go into that -- it's a jungle!" And he certainly was right: the energy values and their use in feeding in the United States is very confusing and disorienting. No one seems to agree with anyone else about which energy value is the appropriate one to use. So to make this as simple and as useful as possible, this book shows TDN, ME, and NEL for dairy cows, and TDN and ME for all other livestock. You can pick the energy value that you're most familiar with for evaluating your pasture and incorporating it into the total ration that you're feeding. See Table 8-4 for pasture classes, as used in New Zealand, expressed in the three energy values more commonly used in the United States.

Regardless of which energy value you pick, you can use the values in the same way that New Zealand farmers do for evaluating pastures and planning the feeding of your livestock (except for milking cows: see dry matter intake discussion below). If you know the energy needs of livestock at different liveweights, production levels, and physiological states, and the energy value of the pasture forage, you can estimate: 1) how much livestock product the pasture will yield, 2) what the dry matter intake per animal should be, 3) forage allowance, 4) how much pasture area is needed, and 5) the stocking rate or carrying capacity of your pastures.

Since you are accustomed to production levels of traditionally mismanaged pastures, the examples of production potentials possible from well managed pastures shown in the feed plan examples at the end of this chapter may seem to be fantastic. But these are examples of real possibilities, based on New Zealand experience with perennial ryegrass-white clover pastures. Although the livestock production potential from other grasses with white clover or other legumes managed well under our conditions are still not known for certain, recent experience by many American farmers indicates that the potential is similar here. Of course, these production levels require excellent management and well developed pastures. You probably won't achieve such high levels of production the first year that you switch from year-round confinement feeding or continuous grazing. Your pastures and soils will need time to repair the damage done from a century or more of abuse. You will need time to develop expertise in Voisin controlled
/ill only come with
ck will need time to
ly if they've been fed
)een bred for years to
probably will need a

OOPS!
p. 204 Table 8-4 should be 8-8
p. 208 Stocking Rate: lb DM/head/day.
p. 210 DM intake = .185 should be .0185
Table 8-5: Item 6. Change 71 to 34
Table 8-6: Item numbers should be 1 to 9.

period of adjustment or selection to achieve more efficient use of protein and energy from pasture forage.

But you can begin now to move toward the production levels shown in the tables. And you might be pleasantly surprised to find that your pasture is not in such terrible shape as you thought. Luckily, nature is very forgiving, otherwise humans long ago would have destroyed their life-support system!

You can take encouragement from the results we got in 1984, when we collected and analyzed 497 pasture forage samples during a 5-month grazing season (May 1-October 1) on six dairy farms in northern Vermont. For one of the farms it was its first season using Voisin controlled grazing management; two were in their second season, two were in their third season, and one was in its fourth season. Even though the farmers had been controlling grazing for such a short time, average forage energy contents on the farms were all within the "good" class during the entire season (Tables 8-3, 8-4). Also, these values were for whole-plant samples. If milking or gaining animals graze paddocks first and get the best forage available, energy levels of that forage probably will be higher than in whole-plant samples. (see Chapter 10 for further discussion of this study)

During 1988 to 1990, 746 forage samples were collected and analyzed from controlled-grazed pastures on farms in Maine, New Hampshire, New York, and Vermont as part of the USDA Low Input Sustainable Agriculture program. All of these analyses (Table 8-3) confirmed the high protein and energy levels possible in well managed pasture forage that we measured in 1984. For comparison in Table 8-3, I separated the results for our farm. We collected forage samples every 2 weeks. During the 1988-1990 sampling period we had been controlling grazing of cattle and sheep on our farm for 7 to 9 years.

Analyze your pasture forage and use the energy values

in management and planning. As you accumulate information and experience, each year your planning will get better, management will improve, production costs will decrease, and your farm's profitability will increase.

TABLE 8-3. Monthly dry matter, crude protein, and energy analyses (dry weight basis) of forage from permanent pastures under Voisin controlled grazing management in Vermont and the Northeast.

Analysis	Month						
	April	May	June	July	Aug	Sept	Oct
Vermont:							
6 farms, 1984 (dry forage yield = 3.7 ton/acre):							
DM, %		22	22	25	24	20	
CP, %		22	20	21	23	25	
TDN, %		70.4	67.8	66.9	70.4	71.3	
ME, Mcal/lb DM		1.16	1.11	1.10	1.16	1.17	
NEL, Mcal/lb DM		.73	.70	.69	.73	.74	
Murphy farm, 1988-90 (dry forage yield = 4.2 ton/acre):							
DM, %	16	19	18	18	18	15	17
CP, %	31	23	25	23	27	30	31
NEL, Mcal/lb DM	.75	.70	.68	.68	.74	.73	.80
Northeast, 1988-90:							
DM, %		19	21	24	22	21	20
CP, %		24	20	22	21	23	24
NEL, Mcal/lb DM		.75	.66	.68	.69	.71	.73

Notes: DM = dry matter, CP = crude protein, TDN = total digestible nutrients, ME = metabolizable energy, NEL = net energy lactation.

Feed Demand: Calculations

Feed demand is expressed as lb DM/acre per day, per

grazing season, or per year, for comparing with feed supply. It is calculated from the energy needed to meet production goals. For milking animals, goals are milk production levels, body condition scores, and birthing date. For growing animals, goals are liveweight gain and total liveweight.

Feed demand is calculated for groups of animals. Since there's a lot of variation among animals, calculate the average goal (e.g. milk production level), and use demand for the average animal.

The energy requirement (energy unit/day) for the average animal of a group to meet a desired production goal (see Tables at end of this Chapter) must be converted to pounds of pasture forage dry matter intake required per day. This is done by dividing energy requirement by the energy content (energy unit/lb DM) of the pasture forage (except for milking cows in the United States: see dry matter intake discussion below).

The diet selected by grazing livestock usually has a higher digestibility and energy content than the total pasture forage, since green leaves are selectively grazed instead of stems or dead material. Intake level or grazing intensity (severity) greatly influences energy content of the forage that is eaten. When forage allowance is low, animals graze to a lower postgrazing mass, and consequently eat more dead material and stems, than animals offered a higher forage allowance. This must be taken into consideration when making feed plans to achieve production goals.

Stocking Rate

Feed demand is defined in lb DM/head/day.d Multiply the individual animal requirement by the stocking rate, to convert to a per acre demand. Stocking rate (animals/acre) can be estimated by:

pasture forage supply ÷ dry matter requirement   or

number of animals ÷ total pasture area

Pasture supply is the total amount of pasture forage estimated that will be produced during the season. Dry matter requirement is the estimated average daily dry matter intake for one animal, multiplied by the estimated number of grazing days.

Dry Matter Intake

From the animal standpoint the amount of dry matter eaten per day depends on a lot of things, including number of bites per minute of grazing time, amount of forage eaten per bite, and total grazing time. Grazing management can influence all of these factors. The main way we can increase dry matter intake is to provide animals with a sward that is in a good physical (pasture mass: 1800-2700 lb DM/acre; short: 4-8 inches tall) and nutritional condition (young vegetative), so that they can graze easily and get a lot of high quality feed per bite.

Dry matter intake seems to be influenced mainly by daily forage allowance. Supplemental feed given to animals decreases their dry matter intake from pasture, but increases their total (supplements + pasture) dry matter intake. Feeding supplements not only decreases pasture dry matter intake, but it increases postgrazing mass, since animals fed supplements are less willing to graze close to ground level.

Given the above complications and animal, sward, and sampling variability, it's no wonder that dry matter intake is the most difficult aspect to determine in livestock grazing pasture. Any estimate of dry matter intake is just that: an estimate. Don't expect it to be precise. Average dry matter intake (lb DM/head/day) can be estimated by:

1) dividing energy requirement by energy content of the pasture forage, as is done in New Zealand (This works fine for all animals except milking cows in the United States. It seems to overestimate how much our cows will eat on pasture, but it may be that we just don't know how much our cows will eat on pasture when they don't receive supplements in the barn, or our pasture forage may contain more energy than shown by our analyses.);

2) using 3 to 3.5% of body weight (some ration balance guides now suggest using 3.99% for lactating cows before and after peak production, and 4.67% at peak production, so maybe our cows do eat as much as New Zealand cows) (use 5.5% of body weight for goats); or

3) using formulas, such as this one from Cornell for milking cows (Although this formula tends to underestimate the amount of dry matter that cows will eat on pasture, it's the one I used in feed plan examples for Holsteins and Jerseys at the end of this chapter, because it may be more familiar and applicable in the USA):

DM intake = (.185 x body wt) + (.305 x fat corrected milk)

Fat corrected milk = (lb milk x % fat x 15) + (lb milk x 0.4)

(A Dairy Pasture Ration Balancer computer program is available at cost from Dr. Ed Rayburn, Seneca Trail Resource Conservation & Development Area, 2 Park Square, Franklinville, NY 14737, telephone: 716/676-5111.)

FEED PROFILE

You can use feed profiles to decide on potential stocking

rates and to balance seasonal feed demand with what you expect the pattern of pasture forage supply to be. Calving and lambing should be timed to synchronize the increased feed demand of lactation with the spring increase in pasture plant growth. Feed profiles concern the average situation, where the total seasonal pasture forage production and pattern of growth is compared with the feed demand for your livestock.

A feed profile indicates when and about how much surplus forage can be machine-harvested and stored, or carried forward as increased pasture cover. It can also provide an estimate of the amount of supplementary feed that may be needed. Extremely low or high pasture cover indicate that you need to do something. Pasture cover can be increased by: 1) changing the timing of events such as calving or lambing to achieve a better balance between supply and demand, 2) increasing the forage allowance by making more pasture area available, or 3) feeding supplements such as hay, green chop, or silage. High average pasture cover (greater than 2200 lb DM/acre) indicates that: 1) you should set aside areas for machine-harvesting surplus forage, 2) bring in more livestock to increase the demand, or 3) change timing of calving or lambing.

Feed profiles can help you to make the right decisions to achieve your goals. Of course, you'll have to monitor pasture cover and livestock performance regularly to know when and if to make management adjustments.

FEED BUDGETS

A feed budget helps you to make management adjustments during the season. It provides information on how best to use available feed to achieve optimum animal production levels and the most profitable use of pasture forage. Since pasture supply and demand usually don't balance the first time that you calculate a feed budget, you have to decide on the least costly way of overcoming a deficit, or on the most profitable way of using surplus forage. When you get the budget to balance, then you can prepare a grazing plan to ration pasture forage to achieve planned dry matter intakes.

Feed budgets can be calculated in different ways, depending on your situation. This is a simple way:

Calculate pasture forage supply:
   present pasture cover (lb DM/acre) +
   pasture growth (lb DM/acre/day x days) -
   pasture cover at end of season (lb DM/acre) =
   pasture supply (lb DM/acre)

Calculate feed demand:
   stocking rate (number of animals/acre/season) x
   daily intake (lb DM/head) x
   number of days in grazing season =
   feed demand (lb DM/acre)

Then compare supply with demand, and calculate the surplus, deficit, or balance. If a deficit occurs, first check the estimated plant growth rates and final pasture cover to make certain that they are correct. If there's still a deficit, you have to evaluate the cost of supplemental feeds, compared to a lower production level from reduced feed intake. If there's surplus forage, you need to consider the effects of increasing feed intake, high postgrazing pasture mass, or of using the excess in some other way, such as machine-harvesting or bringing in more animals.

You should monitor pasture cover to know if it becomes different from what you predicted in your feed budget. If cover does change, you can make the appropriate management adjustment in time.

GRAZING PLANS

Grazing plans are needed when pasture forage is being rationed out and livestock are rotationally grazed. You have to make detailed decisions about how long a group of livestock should stay in each paddock to achieve the planned dry matter intake required to reach your production goal. Before preparing a grazing plan, your feed budget should balance, indicating that there'll be enough pasture forage available, or if and when supplements will be needed.

There are two ways of making grazing plans. The residual method uses relationships among animal production levels, postgrazing pasture mass, and dry matter intake. The forage allowance method uses relationships among animal numbers, pregrazing pasture mass, and forage allowance. Forage allowance (lb DM/head/day) is calculated by dividing pregrazing pasture mass by the stocking density (number of animals/acre in the paddock).

Many graziers prefer the residual method, despite limitations and variability in some production relations, probably because they can see the postgrazing pasture mass and relate to it easier. Forage allowance is a ration, and some people have difficulty relating conceptually to it. Both methods are useful; use the one you like best.

This is how to calculate the number of days a group of livestock should occupy a paddock (occupation period):

Residual method:

$$\text{days} = \frac{(\text{pregrazing - postgrazing mass}) \times \text{paddock area}}{\text{number of animals} \times \text{dry matter intake}}$$

213

Pasture allowance method:

$$\text{days} = \frac{\text{pregrazing pasture mass x paddock area}}{\text{number of animals x forage allowance}}$$

For example, calculate how many days 65, 850-lb Jersey cows producing 55 lb milk/day should occupy a 2-acre paddock having 2200 lb DM/acre good quality pregrazing pasture mass.

Residual method:
Pasture feeding levels for the production goal are a postgrazing pasture mass of 1200 lb DM/acre, and a total intake of 34 lb DM/head/day.

Occupation period = [(2200 - 1200) x 2] ÷ (65 x 34) = 1 day

Forage allowance method:
The pasture feeding level for the production goal is a forage allowance of 2200 lb DM/acre/65 cows x 2 acres = 68 lb DM/head/day. This is a little confusing, because the animals won't eat 68 lb DM/day; they eat 40 to 60% of the forage allowance, leaving the rest as postgrazing mass.

Occupation period = (2200 x 2) ÷ (65 x 68) = 1 day

Both methods give the same answer. In this case, since these are milking cows it would be best to divide the paddock in half with portable fencing, and allow the cows to graze 12 hours in each half.

To make a grazing plan for a rotation, you have to know the pasture mass and area of every paddock. Then list paddocks in the order that they'll be grazed, usually from highest to lowest pasture mass. Also take into consideration location of paddocks within your farm and ease of livestock movement among the paddocks. Then calculate the number of days that animals will occupy each paddock. For long-term plans you'll have to estimate the plant growth that will occur until paddocks are grazed.

Successfully using either method of making a grazing plan depends on the relationships between intake and forage allowance or postgrazing pasture mass. These are influenced by pregrazing mass and sward botanical composition. On short swards having low pregrazing mass, dry matter intake is limited by grazing difficulties, and relationships between plant and animal production and forage allowance are affected. Similar intakes have been measured at the same forage allowance on long swards with pregrazing pasture masses ranging from 1800 to 4500 lb DM/acre, but postgrazing mass increases with increasing pregrazing mass.

When making feed plans, you must consider subsequent effects that your management will have on plant and animal production. For example, swards maintained very short and with low pasture mass will have a high content of green leaves, but slow plant growth rate. In contrast, swards with high postgrazing pasture masses associated with high levels of dry matter intake can result in a lot of rejected forage, reduced forage feeding value, and decreased animal productivity.

FEED PLANNING IN PRACTICE

Feed plans are based on your predictions of future feed supply and demand. You must regularly monitor plant and animal production, and adjust your plan and grazing management accordingly. One of the most variable parts of a feed plan is plant growth rate. If you monitor soil temperature and moisture, this information can help you in predicting plant growth and accumulation. A computer model that predicts pasture plant growth from soil temperature, moisture, and fertility would be helpful, and eventually will be developed as pasture research continues.

You can make feed plans based on predictions of high, low, or average plant growth rate, with management options planned for each level. Such exercises can help you determine when to make important decisions such as feeding supplements or selling livestock, so that the decisions can be made before reaching a critical point.

You can monitor your feed plan by estimating pasture cover at least once a month, and comparing the actual amount with what you predicted. If more detail is needed, check plant growth rates, occupation periods per paddock, and pre- and postgrazing pasture masses every time you move animals during a rotation. If you regularly check milk production levels or weigh animals, you can determine if animal performance goals are being reached.

FEED PLAN EXAMPLES

The examples show total daily dry matter intake needs of livestock. Any supplemental feeding would reduce the amount of forage dry matter eaten on pasture. For instance, if a cow needs to eat 34 lb DM/day, and receives 11 lb DM as silage and grain in the barn at milking, she will only eat 34 - 11 = 23 lb DM/day on pasture.

I. Detailed Feed Profile For A Holstein Dairy Herd

Livestock: 70 synchronized, spring-calving cows with an average bodyweight of 1300 lb. Depending on growing conditions and plant growth, cows graze from April 15 through October 31, or 199 days. Replacement heifers graze earlier and later in season than cows, and follow cows when cows graze. Average milk butterfat content is 3.6%. Average dry matter intake of cows is 42.3 lb DM/head/day.

Pasture: 70 acres; target pasture cover during season will be 1650 lb DM/acre to ensure that forage allowance is adequate to satisfy dry matter intake needs; postgrazing mass should be 1150 to 1350 lb DM/acre.

Goal: produce most of annual milk on pasture.

Feed supply: seasonal average total pasture forage production of 8400 lb DM/acre, with the following pattern of pasture plant growth (lb DM/acre/day), using 2-year average growth rates measured in Vermont:

Month:	April	May	June	July	Aug	Sept	Oct
Growth:	33	59	57	42	40	33	12

Pasture cover on April 15 is 1500 lb DM/acre.

Proposed stocking rate: estimated from average daily dry matter intake (DMI) to obtain total dry matter need per cow for the season. For cows in the United States, it seems best to estimate dry matter intake using 3 to 3.5% of bodyweight, or the Cornell formula discussed above. I used the Cornell formula for the dairy herd examples here, to take into account changes in milk production level. All you have to do is estimate the average amount of milk that a cow will produce per day during a month, average percent butterfat, and calculate daily potential dry matter intake from that. Use those average dry matter intakes to calculate dry matter need for estimating what stocking rate should be during the season.

Stocking rate: pasture supply ÷ DM need = cows/acre

Dry matter need = average daily DMI x grazing days in the season. For this example (using average daily DMI from Table 8-4), it would be:

DM need = 42.3 lb DM/day x 199 days = 8418 lb DM/acre.

Stocking rate = 8400 ÷ 8418 = 1.0 cow/acre

TABLE 8.4. Calculation details of a feed profile for a spring-calving Holstein herd.

	Month						
Item	April	May	June	July	Aug	Sept	Oct
1. Milk prod. lb/day	65	75	75	65	55	55	55
2. DMI, lb DM/day	42.7	45.6	45.6	42.7	39.8	39.8	39.8
3. Cows/acre	1.0	1.0	1.0	1.0	1.0	1.0	1.0
4. Feed demand lb DM/acre/day	42.7	45.6	45.6	42.7	39.8	39.8	39.8
5. Pasture growth lb DM/acre/day	33	59	57	42	40	33	12
6. Change, pasture cover lb DM/acre	-146	415	342	-22	6	-204	-862
7. Pasture cover at end of month lb DM/acre		1354	1769	2111	2089	2095	1891	1029

Notes:

Line 1: pattern of average milk production, lb/cow/day.

Line 2: average daily dry matter intake, calculated with Cornell formula given above.

Line 3: stocking rate, number of cows/acre.

Line 4: average daily DM intake x stocking rate.

Line 6: calculated by subtracting demand (Line 4) from supply (Line 5), and multiplying by the number of days grazing per month.

Line 7. the net change is added to or subtracted from (depending on the sign) the cover present at the end of the previous month.

In the this case (Table 8-4), with a stocking rate of 1.0 cow/acre and cows beginning to graze on April 15, the target pasture cover of at least 1650 lb DM/acre will not be achieved during April and October. Keeping in mind the need to feed newly calved cows generously, a pasture cover that is less than desirable may not be a problem in April, since the plant growth rate becomes very fast during the second half of the month. A lower pasture cover than desired in October also may not be a problem, because the grazing season is ending. Paddocks to be grazed would have to be monitored closely during April and October to be certain that forage allowance is adequate.

An alternative would be to start the cows grazing on about May 1. This would allow more pasture cover to accumulate during April, and consequently during the rest of the season. In this case stocking rate could be increased because of fewer grazing days (8400/42.3 x 184 days = 1.1 cows/acre). If stocking rate is increased (less area used or more animals grazed on same area), pasture cover during October will be less than at the lower stocking rate.

In both cases, surplus forage may be available for other purposes during May and June, according to the calculations below on pasture area needed. Of course, reducing this surplus affects pasture cover during the rest of the season, so be careful how much excess forage is removed.

Pasture area required by month = $\dfrac{\text{DMI x number of cows}}{\text{plant growth rate}}$

May: (45.6 lb DM/d x 70 cows) ÷ 59 lb DM/ac/d = 54 acres
June: (45.6 lb DM/d x 70 cows) ÷ 57 lb DM/ac/d = 56 acres

In October (in this example) other areas could be grazed (e.g. hayland aftermath, or an annual crop seeded to extend the grazing season) to meet the feed deficit and

219

carry the cows through the month, and possibly extend the cow-grazing season beyond October. Another way to continue the cows on pasture would be to provide supplemental feed on pasture, such as hay. This is an example of how to calculate the amount of supplemental feed needed:

Feed requirement = stocking rate x DMI x days grazing

(1.0 cow/acre x 39.8 lb DM/day in October) x 31 days = 1234 lb DM/acre

Pasture forage available = plant growth rate x days + cover at end of previous month - target cover.

In this example: [(12 x 31) + 1891] - 1650 = 613 lb DM/acre

Hay needed = feed requirement - pasture forage available.

Hay needed in this example: 1234 - 613 = 621 lb DM/acre

For this example, hay bales weigh 40 lb, are 85% DM, and have an energy content of 0.56 Mcal NEL/lb DM; in October pasture forage has an energy content of 0.80 Mcal NEL/lb DM.

Pasture DM equivalent per bale = hay bale wt x %DM x hay energy content ÷ pasture energy content

(40 x 0.85 x 0.56) ÷ 0.80 = 24 lb pasture DM equivalent/bale

Bales/acre = hay needed/acre ÷ pasture equivalent/bale

621 ÷ 24 = 26 bales need to be fed per acre.

Balance:

Date	Initial cover	Plant growth	Total pasture forage	Feed need met by hay	pasture
------------------------lb DM/acre------------------------					
Oct 1	1891	372	2263	621	613

Cover at end of October = 1650 lb DM/acre

II. Feed Budget For Above Holstein Dairy Herd

Feed Supply = plant growth rate x days

April:	33 x 30 =	990
May:	59 x 31 =	1829
June:	57 x 30 =	1710
July:	42 x 31 =	1302
Aug:	40 x 31 =	1240
Sept:	33 x 31 =	990
Oct:	12 x 31 =	372
	Total	8433 (round to 8400) lb DM/acre

Feed Demand = stocking rate x average DMI x days

1.0 x 42.3 x 199 = 8418 (round to 8400) lb DM/acre

Balance = feed supply - feed demand
8400 - 8400 = 0

Although the feed profile shows a possible feed deficit in October, in this case the feed budget shows that feed supply and demand will balance. If surplus forage is reduced in May or June, then the deficit in October will have to be filled by feeding supplements or grazing the cows in other areas.

III. Feed Profile For A Jersey Dairy Herd

Livestock: 70 synchronized spring-calving cows with an average bodyweight of 850 lb. Calving timed to match the pattern of pasture production and feed demand. Grazing season for cows usually is from mid-April through October. Heifers begin grazing earlier and graze later, depending on pasture cover. Heifers follow cows after cows start grazing. Average milk butterfat content is 4.9%. Average dry matter intake of cows is 30.8 lb DM/head/day.
Pasture: 50 acres; want pasture cover to be 1650 lb DM/acre to ensure that forage allowance will be adequate; postgrazing mass should be 1150 to 1350 lb DM/acre.

Goal: produce as much of annual milk on pasture as possible.

Feed supply: same as profile for Holstein herd above: 8400 lb DM/acre. Pasture cover on April 15 = 1500 lb DM/acre.

Proposed stocking rate: estimated using Cornell formula for dry matter intake discussed above.

Stocking rate = pasture supply ÷ DM need = cows/acre

For this example, DM requirement would be:
     30.8 lb DM/head/day x 199 days = 6129 lb DM/acre

Stocking rate = 8400 ÷ 6129 = 1.4 cows/acre

TABLE 8-5. Calculation details of a feed profile for a spring-calving Jersey herd

Item	Month						
	April	May	June	July	Aug	Sept	Oct
1. Milk prod. lb/d	45	55	55	45	35	35	35
2. DMI, lb DM/d	31.3	34.8	34.8	31.3	27.8	27.8	27.8
3. Cows/acre	1.4	1.4	1.4	1.4	1.4	1.4	1.4
4. Feed demand lb DM/acre/d	43.8	48.7	48.7	43.8	38.9	38.9	38.9
5. Pasture growth lb DM/acre/d	33	59	57	42	40	33	12
6. Change, pasture cover lb DM/acre	-162	319	249	-56	71	-177	-834
7. Pasture cover at end of month lb DM/acre	1338	1657	1906	1850	1884	1707	873

IV. Feed Budget For Above Jersey Herd

Feed Supply = 8400 lb DM/acre

Feed Demand: 1.4 cow/acre x 30.8 lb DMI/cow x 199 days = 8581 lb DM/acre

Balance: 8400 - 8581 = -181 lb DM/acre deficit

In this case, the budget shows that a feed deficit will occur in October, especially if surplus forage is harvested in May or June. This deficit can be filled in various ways as discussed above for the Holstein herd.

V. Feed Profile For Growing Beef Animals

<u>Livestock</u>: 100 steers; average weight = 550 lb on April 15, when they begin grazing.

<u>Pasture</u>: 80 acres; want pasture cover to be 1300 lb DM/acre during the season to ensure adequate forage allowance. Postgrazing mass should be 1000 lb DM/acre.

<u>Goal</u>: steers to weigh 1000 lb each by November 1. Gains (days x average weight gain/day = lb) will be:

		Start: 550
April:	15 x 2.75 = 41	591
May:	31 x 2.75 = 85	676
June:	30 x 2.75 = 83	759
July:	31 x 2.75 = 85	844
Aug:	31 x 2.20 = 68	912
Sept:	30 x 1.65 = 50	962
Oct:	31 x 1.10 = 34	End: 996

TABLE 8-6. Calculation details of a feed profile for growing beef animals

| Item | Month | | | | | | |
	April	May	June	July	Aug	Sept	Oct
1. lb gain/head/day	2.75	2.75	2.75	2.75	2.20	1.65	1.10
2. Energy req. ME	21.7	26.4	26.4	26.4	26.3	22.6	19.5
3. Forage ME	1.2	1.2	1.2	1.2	1.2	1.2	1.2
4. DMI lb DM/h/day	18.1	22.0	22.0	22.0	21.9	18.8	16.3
5. Steers/acre	1.9	1.9	1.9	1.9	1.9	1.9	1.9
4. Feed demand lb DM/acre/day	34.4	41.8	41.8	41.8	41.6	35.7	31.0
5. Pasture growth lb DM/acre/day	33	59	57	42	40	33	12
6. Change, pasture cover lb DM/acre	-21	533	456	6	-50	-81	-589
7. Pasture cover at end of month lb DM/acre	1479	2012	2468	2474	2424	2343	1754

Notes:

Line 2: energy required in Mcal ME/head/day, (Table 8-10)
for the average daily gains wanted in Line 1.

Line 3: forage energy content, Mcal ME/lb DM, (Table 8-8).

Line 4: dry matter intake/head/day =
required ME/feed ME.

Line 5: stocking rate = feed supply/DMI x grazing days.

VI. Feed Budget For Above Beef Animals

Feed Supply = 8400 lb DM/acre

Feed Demand = stocking rate x DMI x grazing days
1.9 x 20 x 199 = 7562 lb DM/acre

Balance: 8400 - 7550 = 850 lb DM/acre surplus.

The budget indicates that surplus forage will exist. The profile shows that cover will be higher than necessary or desirable during the entire season, unless the surplus is reduced in May or June. The surplus could be reduced by carrying more animals or decreasing the area available to the steers. The area that could be set aside and used for other purposes is calculated as follows:

$$\text{Area required} = \frac{\text{DMI (lb DM/head/day x number of cows)}}{\text{plant growth rate}}$$

For May: 22 x 100/59 = 37 acres needed
For June: 22 x 100/57 = 39 acres needed

So 41 to 43 acres could be used for grazing other livestock or machine harvesting in May or June. If surplus forage is reduced, the feed profile will change during the rest of the season. Pasture cover should be monitored closely to make certain that the target cover of 1300 lb DM/acre is maintained.

VII. Feed Profile For Ewes And Lambs

Livestock: 300 ewes, lambing April 1 with a 1.7 lambing rate that results in 510 lambs. Begin grazing April 10.

Pasture: 50 acres; want pasture cover to be 1300 lb DM/acre during the season to ensure adequate forage allowance. Postgrazing mass should be 1000 lb DM/acre. Pasture cover is 1000 lb DM/acre on April 1, so by April 10 cover will be:

10 days x 33 lb DM/day = 1330 lb DM/acre.

<u>Goal</u>: lambs to reach 105-lb market weight by October 31 on pasture alone. Lamb average liveweight gains should be:

	Birthweight:	10 lb
April: 30 x .52 = 16		26
May: 31 x .52 = 16		42
June: 30 x .52 = 16		58
July: 15 x .52 = 8		66
Weaned July 15		
July: 16 x .44 = 7		73
Aug: 31 x .44 = 14		87
Sept: 30 x .44 = 13		100
Oct: 31 x .22 = 7	End:	107 lb

<u>Stocking rate</u>: in this case is estimated by dividing the number of animals by acres of pasture:

Ewes : 300 ÷ 50 = 6 ewes/acre

Lambs : (7% reach market weight by weaning and are sold; 20% of remaining lambs reach market weight by August 31 and are sold; 20% of remaining lambs reach market weight by September 30 and are sold; all remaining lambs are sold on November 1):

April-July 15: 1.7 x 300 ewes = 510 lambs
510 ÷ 50 acres = 10.2 lambs/acre

July 16-August 31: 93% x 510 lambs = 474 lambs
474 ÷ 50 acres = 9.5 lambs/acre

September 1-30: 80% x 474 lambs = 379 lambs
379 ÷ 50 acres = 7.6 lambs/acre

October 1-31: 80% x 379 lambs = 303 lambs
303 ÷ 50 acres = 6.1 lambs/acre

TABLE 8-7. Calculation details of a feed profile for ewes and lambs (lambs weaned July 15).

	Month						
Item	April	May	June	July*	Aug	Sept	Oct
Ewes							
1. Energy need, ME	7.9	6.9	5.5	3.9	2.2	2.2	2.2
2. Forage ME	1.2	1.2	1.2	1.2	1.2	1.2	1.2
3. DMI/acre	39.5	34.5	27.5	19.5	11.0	11.0	11.0
Lambs							
4. Energy need, ME	0.5	1.0	1.9	3.0	4.6	4.6	4.6
5. Forage ME	1.2	1.2	1.2	1.2	1.2	1.2	1.2
6. DMI/acre	4.3	8.5	16.2	24.8	36.5	29.1	23.4
7. Total demand	43.8	43.0	43.7	44.3	47.5	40.1	34.4
8. Pasture growth lb DM/acre/day	33	59	57	42	40	33	12
9. Change, pasture cover lb DM/acre	-216	496	399	-71	-233	-213	-694
10. Pasture cover at end of month lb DM/acre		1114	1610	2009	1938	1705	1492	798

Notes:

Lines 1 & 4: energy required in Mcal ME/head/day, from Tables 8-11, 12, and 13.

Lines 2 & 5: forage energy, Mcal ME/lb DM (Table 8-8).

Lines 3 & 6: dry matter intake/acre/day = (required ME/feed ME) x stocking rate/acre.

Line 7: total DM demand = line 3 + line 6.

July*: values shown are averages because of lambs being weaned on July 15; ewe daily energy needs 7/1-15 = 5.5 Mcal ME, and 7/16-31 = 2.2 Mcal ME; lamb energy needs 7/1-15 = 1.9 Mcal ME, and 7/16-31 = 4.1 Mcal ME.

VIII. Feed Budget For Above Ewes And Lambs

Feed supply: 8400 lb DM/acre

Feed demand: stocking rate x average DMI x days
(In this case, DMI in the profile is expressed per acre and includes stocking rate.)

Ewes
4/10-7/15: 32.3 lb DM/h/d x 97 days = 3133 lb DM
7/16-10/31:   11.0 x 108 = 1188

Lambs
4/10-715   11.3 x 97 = 1096
7/16-8/31   34.5 x 47 = 1622
9/1-9/30   29.1 x 30 = 873
10/1-10/31   23.4 x 31 = 725
                        Total   8637

        Supply 8400
        Demand 8637
        Deficit   237 lb DM/acre

This deficit could be covered by grazing the dry ewes on other areas, such as on land needing improvement, hayland aftermath, or an annual crop seeded for this purpose.

ENERGY/PRODUCTION TABLES

The tables present estimates and examples of potential production levels on pasture alone, without supplements. Higher production levels and pasture stocking rates or carrying capacities are possible if some supplements are fed.

Energy values in the tables are based on forage dry matter (DM) and are expressed as percent (in Table 8-8 only) or pounds of total digestible nutrients (TDN), Mcal of metabolizable energy (ME) per pound of forage DM, and Mcal of net energy lactation (NEL) per pound of forage DM.

The energy values in the tables are the amounts of energy needed by animals and provided by pasture forage. So the values can be looked at in two ways:

1. For a given level of production, an animal of a given weight and physiological state needs a certain amount of daily energy, which can be obtained from certain pasture classes, based on their available energy contents.

2. A given pasture class, based on its available energy content, can support certain production levels of certain animal weights and physiological states.

TABLE 8-8. Pasture classification based on forage energy content, (dry matter basis) expressed as TDN, ME, and NEL as used in the United States, compared to ME as used in New Zealand.

Class	New Zealand ME	United States energy values		
		TDN	ME	NEL
	Mcal/kg	%	----Mcal/lb----	
Poor	1.8	49.8	0.8	0.50
Fair	2.2	60.9	1.0	0.62
Good	2.6	71.9	1.2	0.75
Excellent	3.0	83.0	1.4	0.87

Note: energy values are range midpoints.

TABLE 8-9. Daily energy (TDN) needs of milking cows, according to liveweight (lb), milk and butterfat (%BF) production, and pasture class.

Live-weight & %BF	Pas-ture class	Milk production, lb/day							
		0	15	25	35	45	55	65	75
		----------------------TDN----------------------							
	Poor	7.3	16.3	----	----	----	----	----	----
800	Fair	7.1	14.6	20.7	----	----	----	----	----
4.9%	Good	6.8	13.7	18.8	24.4	31.1	----	----	----
	Exc.	6.6	13.5	18.1	23.2	28.7	34.7	----	----
	Poor	9.3	16.5	22.7	----	----	----	----	----
1100	Fair	9.0	15.2	20.1	25.4	----	----	----	----
3.8%	Good	8.6	14.6	18.5	22.9	27.7	29.1	----	----
	Exc.	8.4	12.1	16.1	20.2	24.5	28.8	35.7	----
	Poor	10.4	17.4	22.7	----	----	----	----	----
1300	Fair	10.1	16.1	20.5	25.4	30.9	----	----	----
3.6%	Good	9.7	15.2	19.2	23.4	27.9	32.8	38.3	----
	Exc.	9.3	14.8	18.7	22.5	26.8	31.2	35.3	40.1

TABLE 8-9. (continued) Daily energy (ME, NEL) needs of milking cows.

Live-weight & %BF	Pasture class	Milk production, lb/day							
		0	15	25	35	45	55	65	75
		----------------------------ME----------------------------							
	Poor	12.7	28.0	----	----	----	----	----	----
800	Fair	12.1	25.4	35.7	----	----	----	----	----
4.9%	Good	11.7	23.6	32.5	42.3	53.8	----	----	----
	Exc.	11.2	23.2	31.4	40.2	49.6	60.1	----	----
	Poor	16.1	28.4	39.0	----	----	----	----	----
1100	Fair	15.4	26.2	34.6	43.7	----	----	----	----
3.8%	Good	14.8	24.9	32.0	39.7	48.0	50.3	----	----
	Exc.	14.3	20.9	27.8	35.0	42.4	49.8	61.7	----
	Poor	18.3	30.2	39.7	----	----	----	----	----
1300	Fair	17.6	28.0	35.7	44.1	53.8	----	----	----
3.6%	Good	16.8	26.5	33.5	40.1	48.2	56.9	66.3	----
	Exc.	16.3	25.8	32.6	39.2	46.3	54.0	65.9	69.4
		----------------------------NEL----------------------------							
	Poor	7.5	16.8	----	----	----	----	----	----
800	Fair	7.3	15.2	21.4	----	----	----	----	----
4.9%	Good	7.0	14.1	19.5	25.4	32.3	----	----	----
	Exc.	6.8	13.9	18.8	24.1	29.8	36.1	----	----
	Poor	9.7	17.4	23.8	----	----	----	----	----
1100	Fair	9.5	16.1	21.2	26.7	----	----	----	----
3.8%	Good	8.9	15.2	19.4	23.8	28.8	30.2	----	----
	Exc.	8.8	12.8	17.0	21.0	25.4	29.8	37.0	----
	Poor	10.6	17.4	22.9	----	----	----	----	----
1300	Fair	10.1	16.3	20.7	25.6	31.3	----	----	----
3.6%	Good	8.4	15.4	19.6	23.6	28.9	34.1	39.8	----
	Exc.	8.3	15.0	19.4	22.7	27.8	32.4	35.7	41.6

TABLE 8-10. Daily energy (TDN) needs of growing beef cattle or dairy heifers, according to liveweight, pasture class, & weight gain.

Class	Weight gain, lb/day						
	0	0.55	1.10	1.65	2.20	2.75	3.30
	----------------------------TDN-----------------------						
Liveweight = 450 lb							
Poor	6.5	7.9	9.7	11.8	----	----	----
Fair	6.2	7.3	8.7	10.2	12.3	----	----
Good	6.0	6.9	8.0	9.2	10.8	12.5	----
Excellent	5.8	6.5	7.5	8.5	9.8	11.2	13.0
Liveweight = 650 lb							
Poor	7.9	9.6	11.5	14.1	----	----	----
Fair	7.7	9.0	10.5	12.3	14.7	----	----
Good	7.3	8.4	9.6	11.0	12.9	15.3	----
Excellent	7.1	8.0	9.1	10.2	11.7	13.5	15.8
Liveweight = 850 lb							
Poor	9.3	11.3	13.5	16.6	----	----	----
Fair	9.0	10.6	12.3	14.5	17.2	----	----
Good	8.7	9.9	11.3	13.1	15.2	18.0	----
Excellent	8.3	9.5	10.6	12.0	13.8	16.0	18.8

TABLE 8-10. (continued). Daily energy (ME) needs of growing beef cattle or dairy heifers, according to liveweight, pasture class, & weight gain.

			Weight gain, lb/day				
Class	0	0.55	1.10	1.65	2.20	2.75	3.30
				ME			
Liveweight = 450 lb							
Poor	11.2	13.7	16.8	20.5	----	----	----
Fair	10.8	12.7	15.0	17.7	21.2	----	----
Good	10.4	12.0	13.8	16.0	18.6	21.7	----
Excellent	10.0	11.3	12.9	14.7	16.9	19.4	22.5
Liveweight = 650 lb							
Poor	13.7	16.6	19.9	24.4	----	----	----
Fair	13.3	15.5	18.1	21.2	25.4	----	----
Good	12.7	14.6	16.6	19.1	22.4	26.4	----
Excellent	12.2	13.8	15.7	17.7	20.3	23.4	27.3
Liveweight = 850 lb							
Poor	16.1	19.6	23.4	28.7	----	----	----
Fair	15.6	18.3	21.3	25.1	29.8	----	----
Good	15.0	17.2	19.5	22.6	26.3	31.1	----
Excellent	14.4	16.4	18.3	20.8	23.8	27.7	32.5

TABLE 8-11. Daily energy (ME) needed for ewes and lambs during lactation when grazing good quality pasture.

Ewe	Period of lactation		
liveweight	Early	Mid	Late
90	6.2	5.5	4.8
120	7.9	6.9	5.5
150	8.6	7.6	6.2
Lamb pasture need	0.5	1.0	1.9

TABLE 8-12. Daily energy (TDN or ME) needed for maintenance of ewes of different liveweights (lb), according to pasture class.

Live-weight	Pasture class							
	Poor		Fair		Good		Excellent	
	TDN	ME	TDN	ME	TDN	ME	TDN	ME
90	1.2	2.1	1.2	2.0	1.1	1.9	1.0	1.8
120	1.4	2.4	1.3	2.3	1.3	2.2	1.2	2.1
150	1.5	2.6	1.4	2.5	1.4	2.4	1.3	2.3

TABLE 8-13. Daily energy (TDN or ME) needs of weaned lambs, according to liveweight (lb), pasture class, and weight gain.

Live wt.	Pasture class	Weight gain, lb/day									
		0		0.11		0.22		0.44		0.66	
		TDN	ME	TDN	ME	TDN	ME	TDN	ME	TDN	ME
	Poor	0.8	1.3	1.2	2.1	---	---	---	---	---	---
45	Fair	0.8	1.3	1.0	1.8	1.4	2.4	---	---	---	---
	Good	0.7	1.2	1.0	1.7	1.3	2.2	2.0	3.5	---	---
	Exc.	0.6	1.1	0.9	1.6	1.2	2.0	1.7	3.0	2.5	4.4
	Poor	1.0	1.7	1.4	2.5	---	---	---	---	---	---
65	Fair	1.0	1.7	1.3	2.3	1.7	3.0	---	---	---	---
	Good	0.9	1.6	1.2	2.1	1.6	2.7	2.4	4.1	---	---
	Exc.	0.9	1.5	1.2	2.0	1.4	2.5	2.1	3.6	2.9	5.0
	Poor	1.2	2.1	1.7	3.0	---	---	---	---	---	---
85	Fair	1.2	2.0	1.6	2.8	2.0	3.5	---	---	---	---
	Good	1.1	1.9	1.4	2.5	1.8	3.1	2.7	4.6	---	---
	Exc.	1.0	1.8	1.3	2.3	1.7	2.9	2.3	4.0	3.2	5.5

TABLE 8-14. Daily energy (TDN or ME) needs of goats, according to liveweight (lb), weight gain (lb/day), and physiological state.

Physiological state and liveweight	Weight gain	Energy need	
		TDN	ME
Wethers and dry does			
50	0-0.15	1.4	2.4
60	0-0.10	1.4	2.5
80	0-0.05	1.6	2.8
100	0	1.7	3.0
120	0	1.7	3.0
Pregnant does (last 8 weeks)			
50	0.38	2.0	3.4
60	0.35	2.0	3.5
80	0.33	2.2	3.8
100	0.30	2.4	4.1
Nursing does			
50	-0.05-0	2.1	3.6
60	-0.05-0	2.2	3.8
80	-0.05-0	2.4	4.1
100	-0.05-0	2.5	4.3
Growing kids and yearlings			
20	0.30	1.2	2.0
40	0.25	1.4	2.5
60	0.20	1.7	3.0
80	0.10	1.8	3.1
Developing bucks			
80	0.30	2.3	4.0
100	0.20	2.4	4.1
120	0.10	2.3	3.9

TABLE 8-15. Daily energy (TDN or ME) needs of milking goats per 2 pounds (1 quart) of milk produced, according to milk butterfat content (BF%).

Milk BF%	Energy need TDN	ME
3.0	0.69	1.19
3.5	0.73	1.27
4.0	0.79	1.36
4.5	0.84	1.45
5.0	0.89	1.54
5.5	0.94	1.63
6.0	0.99	1.72

TABLE 8-16. Daily energy (TDN or ME) needs of mature horses, pregnant mares, and lactating mares, according to degree of work and liveweight (lb).

Degree of work or physiological state	Energy needed at liveweights of 450		900		1100		1300	
	TDN	ME	TDN	ME	TDN	ME	TDN	ME
Horse at rest (maintenance)	3.9	6.7	6.6	11.4	7.7	13.4	8.9	15.4
Horse doing light work (2 hrs/day)	7.3	12.7	8.7	15.1	10.4	18.0	12.0	20.8
Horse doing medium work (4 hrs/day)	6.2	10.8	11.3	19.5	13.6	23.5	16.0	27.6
Mare, late gestation	4.1	7.1	7.1	12.2	8.3	14.3	9.5	16.4
peak lactation	7.2	12.5	11.6	20.0	13.1	22.6	14.2	24.6

TABLE 8-17. Daily energy (TDN or ME) needs of growing horses, according to final mature weight (lb), age (mo), liveweight (lb), and weight gain (lb/day).

Mature weight & age	Live-weight	Weight gain	Energy need TDN	ME
450-lb mature weight				
3	100	1.54	3.5	6.1
6	200	1.10	4.0	7.0
12	300	0.44	3.8	6.6
18	350	0.22	3.8	6.6
42	450	0.00	3.9	6.7
900-lb mature weight				
3	200	2.20	4.9	8.5
6	375	1.43	5.6	10.2
12	575	0.88	6.5	11.2
18	725	0.55	6.7	11.6
42	900	0.00	6.6	11.4
1000-lb mature weight				
3	250	2.43	5.7	9.9
6	500	1.76	7.3	12.6
12	700	1.21	8.0	13.8
18	900	0.77	8.1	14.1
42	1000	0.00	7.7	13.4
1300-lb mature weight				
3	300	2.76	6.7	11.6
6	600	1.87	8.1	14.1
12	850	1.32	9.0	15.5
18	1000	0.77	9.1	15.7
42	1300	0.00	8.9	15.4

TABLE 8-18. Daily energy (TDN or ME) needs of growing and breeding pigs, according to liveweight (lb) and physiological state.

Liveweight & physiological state	Energy need TDN	ME	Liveweight & physiological state	Energy need TDN	ME
Growing pigs			Lactating gilts		
10-20	1.2	2.0	300-450	9.1	15.8
20-40	2.4	4.2			
45-80	3.1	5.4	Lactating sows		
80-130	4.6	7.9	450-550	10.0	17.4
130-220	6.4	11.1			
			Young boars		
Bred gilts			250-400	4.6	7.9
240-350	3.6	6.3			
			Adult boars		
Bred sows			400-550	3.6	6.3
350-550	3.6	6.3			

TABLE 8-19. Daily energy (TDN or ME) needs of chickens, according to physiological state and age.

| Physiological | Energy need | |
state & age	TDN	ME
Broilers		
0-3 weeks	0.82	1.42
3-6 weeks	0.84	1.46
6-9 weeks	0.86	1.48
Replacement pullet layers		
0-6 weeks	0.77	1.34
6-12 weeks	0.79	1.36
12-18 weeks	0.79	1.36
18 weeds-laying	0.79	1.37
Layers	0.75	1.30

TABLE 8-20. Daily energy (TDN or ME) needs of growing turkeys, according to sex and age (weeks).

Sex & age	Energy need	
	TDN	ME
Male		
0-4	0.73	1.27
4-8	0.76	1.32
8-12	0.79	1.36
12-16	0.81	1.41
16-20	0.84	1.45
20-24	0.87	1.50
Female		
0-4	0.73	1.27
4-8	0.76	1.32
8-11	0.79	1.36
11-14	0.81	1.41
14-17	0.84	1.45
17-20	0.87	1.50

TABLE 8-21. Daily energy (TDN or ME) needs of ducks and geese, according to physiological state.

Physiological state	Energy need	
	TDN	ME
Ducks		
Starting	0.76	1.32
Growing	0.76	1.32
Breeding	0.76	1.32
Geese		
Starting	0.76	1.32
Growing	0.76	1.32
Breeding	0.76	1.32

CHAPTER 8

REFERENCES

Chase, L.E., and C.J. Sniffen. 1985. Equations used in ANALFEED, a Visicalc template. Cornell Dept. Animal Science. Ithaca, New York. Mimeo.

Ensminger, M.E., and C.G. Olentine, Jr. 1978. *Feeds and Nutrition -- Complete.* Ensminger Pub. Co., Clovis, California. 1417 p.

Flanagan, J.P., J.P. Hanrahan, and P. O'Malley. 1987. *Lowland Sheep Production - Blindwell System.* An Foras Taluntais. Western Research Centre, Belclare, Tuam, Ireland. Sheep Series No. 2. 20 P.

Holmes, C.W. 1987. Pastures for dairy cows. p. 133-143. In. A.M. Nicol (ed.) *Feeding Livestock on Pasture.* New Zealand Society of Animal Production. Hamilton. Occasional Publication No. 10.

Holmes, C.W., and G.F. Wilson. 1984. *Milk Production From Pasture.* Buttersworth of New Zealand. Wellington. 319 p.

Milligan, K.E., I.M. Brookes, and K.F. Thompson. 1987. Feed planning on pasture. p. 75-88. In. A.M. Nicol (ed.) *Feeding Livestock on Pasture.* New Zealand Society of Animal Production. Hamilton. Occasional Publication No. 10.

Rayburn, E. 1990. *Forage Quality of Intensive Rotationally Grazed Pastures - 1989.* Seneca Trail Resource Conservation and Development Area and Cornell Extension Service. Franklin, New York. 34 p.

9

Extending The Grazing Season

> On the dead branch
> A crow settles --
> Autumn evening
>   Basho

The grazing season can be extended at either end or during the season when stored feed would otherwise have to be fed because of a shortage of pasture forage. The best way to extend your grazing season is to apply Voisin controlled grazing management. This alone will add months of additional grazing time at usual season ends, and will minimize or eliminate plant production slumps that otherwise occur during the season. Other ways of doing it range from carrying deferred pasture forage over winter for early spring grazing, to seeding forage chicory or brassicas for autumn and late fall grazing. Using specially seeded crops to make up for pasture shortages incur costs of seedbed preparation, fertilizer, seed, fuel, machinery, labor, and time. So before seeding a crop, compare estimated costs with projected returns to determine if what you're thinking of doing will pay for your trouble. Ask the Extension Service for suggestions about cultivars, seeding rates, and times suited to your conditions. Give animals small paddocks or narrow strips of these crops with portable fencing, so that most of the valuable forage is eaten. Annual crops for fall grazing should only be seeded on land that won't be at risk of soil erosion during winter after the cover has been grazed off.

AUTUMN

Autumn to early winter is the time that I think of first in trying to extend the grazing season. It's the only time in Vermont when we can or need to extend the grazing season, because in early spring (March to mid-April) we have what is called Mud Season. There're many Vermont jokes about it, and schools still regularly schedule a Mud Season break. It goes back to the time before road pavement when people simply couldn't get to school from where they were in early spring. We still can barely get through our gravel road, let alone try to put livestock on pasture during Mud Season! Of course fall and summer also are cool and wet in Vermont, but it's the degree of wetness that matters.

Hay Aftermath

Besides improving grazing management, the next easiest and cheapest way of extending the grazing season is to graze autumn regrowth of hayland, rather than machine-harvesting and feeding it as green-chop or storing it. This is very simple to do with portable fencing.

At this time plant regrowth rate will be too slow to worry about regrazing before adequate recovery. But it's still best to provide just enough forage allowance for short grazing periods, by using small paddocks rather than having livestock wandering around large hay fields.

Small paddock subdivisions save animal energy, reduce plant crown damage (especially alfalfa), soil poaching (punching holes), and soil compaction by restricting animal movement. The higher stocking density within paddocks, compared to using large fields, results in quick, more complete grazing of available forage. The combination of these effects from small paddock grazing results in higher production per acre, compared to the grazing of large areas. There shouldn't be any rejected forage, because there were no previous manure droppings.

Deferred Or Stockpiled Pasture

Pasture surplus forage can be deferred from grazing (stockpiled) and carried forward into fall or into any other time of the season when there's usually a shortage. This can be done by grazing all or parts of your pastureland just enough (high postgrazing pasture mass) to keep grasses from flowering during times of rapid plant growth. Then set aside less intensively grazed parts for grazing in the fall or during other periods of slow plant growth. This really is the only way to deal with surplus forage if your pasture is all on rough land where you can't use machinery. It's also a way to avoid having to use machinery even if the topography of the land would allow it. In using this method try to carry forward different parts of your pasture each year, so that the sward doesn't thin out from less grass tillering or loss of low-growing plants due to shading in consistently high pasture masses.

Stockpiling pasture forage works especially well with tall fescue. Freezing temperatures of fall cause changes in fescue plants that make them more palatable to livestock.

Crop Residues

Farmers traditionally graze corn fields after harvest with

beef cattle to eat any of the crop that was missed. There's no reason why beef and other livestock can't graze all crop residues, including those of small grains. When properly supplemented, crop residues can form a valuable and inexpensive part of the ration. Plant parts left in the field can be almost as important nutritionally as those removed with a combine. For example, small grain (barley, oats, wheat) straw contains about 5 percent protein and 37 percent energy. Grazing residues in the field helps decrease weed populations, and returns much of the crop organic matter to the soil. It certainly is a better solution than burning to remove unwanted residues. Check labels of pesticides that were used on the crop to be certain that the residue can be grazed.

Small Grains

Small grains such as wheat, oats, rye, barley, and triticale can be seeded in early autumn to provide good grazing 6 to 8 weeks later. Because small grains grow best under cool temperatures, they can be drilled into pasture sod in late summer or early fall. At this time pasture plant growth slows down so there is little competition with the small grain seedlings. No herbicides are needed. As paddocks are grazed in the last rotation they can be seeded to small grains. For example, a farmer in Virginia drills about 55 lb/acre of wheat or rye with 15 lb/acre of Austrian winter pea, or 50 lb/acre of oats with 15 lb/acre of Green Globe turnip seed into pasture sod in early fall. These seedings provide excellent grazing during December and January.

Successful use of small grains drilled into and grazed on pasture sod in the fall and early winter depends on having a well drained soil. If the soil is too wet, the pasture may become badly damaged and consequently will not be very productive the following grazing season.

Forage Brassicas And Chicory

Forage chicory and brassicas such as turnips, rape, kale, and swedes can be very useful in cutting feed and feeding costs during mid to late season. These cold-hardy, fast-growing species establish easily, thrive under many conditions, crowd out weeds, and can be grazed well into winter in many areas. Seed costs are low ($3 to $8/acre), and only moderate soil fertility is needed. Since dry matter production is high (4 ton DM/acre for turnips and rape; 6 ton DM/acre for kale and swedes), 1 acre of these plants generally will feed 170 cows or 1000 sheep for a day. They can be seeded anytime from late spring to midsummer, and be ready to graze in 6 weeks from seeding. Rape and chicory can be grazed in autumn and again later in the fall or early winter. Chicory is a perennial, so it can provide grazing in early spring, and at other times of the season.

Annual Ryegrass

Tetraploid annual ryegrass can be seeded with an oat crop for fall grazing. Just seed 15 to 20 pounds of ryegrass per acre in the spring along with oats seeded at the normal seeding rate of about 3 bushels per acre. The ryegrass seed can be mixed with the oats in the seeder. Harvest the oats for grain and straw in July or August. Then begin grazing the ryegrass in September when it's about 6 inches tall. Ration out the feed and keep animals from grazing regrowth. Depending on the weather at your location, the ryegrass will regrow for successive grazings.

MIDSUMMER

Warm-Season Perennial Grasses

Warm-season native grasses such as switchgrass, big

bluestem, and Indiangrass can complement cool-season species such as Kentucky bluegrass. Cool-season grasses produce most of their forage by early to mid-June, just when warm-season grasses are reaching their period of greatest productivity.

To manage them correctly, warm-season grasses must be grown separately from cool-season species. Iowa and Missouri farmers have found that it works well to keep 25 to 40 percent of their pasture area in warm-season grasses, and the rest in cool-season species.

Warm-season grasses provide about a month of grazing in midseason, and another grazing after they have regrown. The amount of grazing time from warm-season grasses can be extended by seeding species separately that reach top production levels at different times, probably because of varying temperature and light requirements. For example, in Missouri switchgrass begins producing a lot of forage in late May to early June, big bluestem reaches peak production in July, followed by Indiangrass in August to early September.

Warm-season grasses should be grazed when they are 15 to 18 inches tall, and only grazed down to about 6 inches tall. They don't recover well if grazed off close to the ground, and will persist only if allowed to recover adequately after their midsummer productive period.

Manure or nitrogen fertilizer should be applied to warm-season grass swards, because they are so dense and tall that they choke out most legumes.

Warm-Season Annual Grasses

If your pasture production usually declines so much in midsummer that you need supplemental feed, you can seed annual grasses such as sudangrass, forage sorghum, and pearlmillet in advance to provide grazing during that time. These grasses grow rapidly, withstand considerable moisture stress, and produce high total forage yields of 3 to 6 ton DM/acre.

The highest forage yields are gotten from these grasses if they're allowed to grow to 30 to 48 inches tall before grazing. At that height only very narrow strips of the forage can be offered to livestock to minimize waste. Most uniform paddock grazing with least waste is achieved by grazing these grasses when they're 8 to 12 inches tall. Regrowth of summer annual grasses depends greatly on amount of leaf surface, and presence and growth of meristematic buds left on the stubble. So don't graze these grasses lower than 4 to 6 inches from the soil surface.

Be careful when grazing sudangrass and sorghum-sudangrass hybrids so that your livestock don't become poisoned by hydrocyanic acid that may form in these plants. Hydrocyanic acid potential differs with cultivar, stage of growth, level of nitrogen fertilization, and environmental conditions. Hybrids tend to form hydrocyanic acid more than sudangrass. Higher concentrations of hydrocyanic acid develop in new growth of grass, at high levels of nitrogen fertilization, and under stress conditions of drought or frost. So apply only low amounts of nitrogen fertilizer for these grasses, and don't graze them at early growth stages or right after a drought or frost.

EARLY SPRING

Grazing of pasture or autumn-seeded small grains can be deferred in the fall so that the forage can be carried over winter for early spring grazing. Deferred forage maintains nutritional quality, especially if it's overwintered under snow and freezing temperatures. Winter wheat will regrow after fall grazing, and provide forage for early spring grazing.

REFERENCES

Brown, C.S., and J.E. Baylor. 1973. Hay and pasture seedings for the Northeast. p. 437-447. In M.E. Heath (ed.) *Forages*. Iowa State University Press. Ames.

Cramer, C. 1989. 12 ways to make pastures really produce. The New Farm. May/June. p. 25-29.

Cramer, C. 1989. "Offbeat" feeds save 20-35%. The New Farm. September/October. p. 30-33.

Cramer, C. 1990. "Grass farming" beats corn! The New Farm. September/October. p. 10-16.

Fribourg, H.A. 1973. Summer annual grasses and cereals for forage. p. 344-357. In M.E. Heath (ed.) *Forages*. Iowa State University Press. Ames.

Jung, G.A. et al. 1987. Grazing systems, herbage quality, and animal behavior on warm-season and cool-season pastures on hill country in NE USA. p. 45-63. In F.P. Horn (ed.) *Grazing Lands Research at the Plant-Animal Interface*. Winrock International. Morrilton, Arkansas.

Knight, A. 1988. Making milk for $2.50. American Agriculturalist. November. p. 15-16.

McIver, D. 1991. Quadruple your stocking rate. The New Farm. May/June. p. 17-21.

Reid, R.L., and G.A. Jung. 1973. Forage-animal stresses. p. 639-53. In M.E. Heath (ed.) *Forages*. Iowa State Univ. Press.

Zahradnik, F. 1986. Bridge the midsummer pasture gap. The New Farm. January. p. 10-13.

10

Economics Of Feeding Livestock On Pasture

> I scooped up the moon
> In my water bucket...
> And spilled it on the grass
>   Ryuho

Not so very long ago economic studies concerning the feeding of livestock on pasture under controlled grazing management in the United States were rare. Now they can be found more easily, ranging from individual farm case studies to linear programming models that apply to many farms. The studies invariably show that farmers who use Voisin controlled grazing management receive benefits that significantly exceed additional costs of adopting the method.

Costs include fencing as one of the first considerations when converting to controlled grazing. Many farms already have perimeter fences in place, and only paddock subdivision are needed. Sometimes only portable fencing needs to be purchased. Since benefits and costs are compared on an annual basis, fencing costs are amortized over their estimated useful lives to convert to an annual cost basis. Costs of providing drinking water to grazing livestock also are amortized to convert to an annual basis. Costs of moving animals and portable fencing, and providing water are based on the amount of time needed to do the work. This labor is given an arbitrary value per hour to include it as a cost. In some situations additional costs may exist, such as seeding, fertilizing, or mowing.

Benefits of converting to controlled grazing reflect the improved quantity and quality of forage produced from pasture. This results in purchasing less grain and hay. Since less grain and hay need to be fed, the amount of labor is reduced. It takes two to three times as much labor to feed the large amount of hay and grain required by confined animals as it does to feed the reduced amount needed by animals under controlled grazing. Other benefits include less labor for manure handling and decreased bedding expense, since animals are confined less under controlled grazing. Since less crops have to planted, fertilized, cared for in controlling pests, harvested, stored, and fed for animals under controlled grazing, less labor is needed in these activities. It follows that less machinery use and repair, fuel, seed, fertilizer, pesticides, and electrical power are needed with decreased cropping. With decreased cropping there would also be less soil erosion and water pollution, but our society still doesn't place an economic value on these, so they aren't included in economic analyses.

Other benefits may not be immediately quantifiable, such as improved herd management due to closer observation of animals when moving them in the pasture. Herd health generally improves. Both of these aspects can save veterinary expenses, and result in improved conception, production, and longevity of livestock. On dairy farms, milk quality may improve by grazing cows on well managed pasture, and result in premiums received for the milk.

Because dairy farmers have been experiencing severe financial problems lately, most of the studies that have been done concern dairy cow feeding on pasture. All of the studies show the same thing: it pays to feed livestock on well managed pasture as much and for as long as possible during the year. Some of the studies are presented here to prove the point.

DAIRY

Six Vermont Farms

This study is included mainly for historical interest. Other studies done since then have shown greater benefits to farmers. Between May 1 and October 1 1984, we (John Rice, David Dugdale, and I) sampled forage from permanent pastures that were being grazed under Voisin controlled grazing management on six northern Vermont dairy farms. We did this to provide farmers with current estimates of feeding value of forage that their cows were grazing, so that they could balance their rations accordingly throughout the season, and determine if the profitability of their farms improved by using controlled grazing management.

Samples were taken each time cows were about to enter a paddock. We took a total of 497 samples and analyzed them chemically and with near infrared reflectance spectroscopy. (Table 10-1)

The analyses showed that forage from pastures under controlled grazing management had high quality during the entire season, especially in terms of protein and energy. If considered according to New Zealand criteria, crude protein content of the forage was sufficient, and no more should have been needed in balancing rations. But, for the reasons mentioned above, cows producing over 55 pounds of milk per day were fed protein concentrates.

253

TABLE 10-1. Monthly analyses (dry weight basis) of forage from permanent pastures grazed with Voisin controlled grazing management on six Vermont dairy farms from May 1 to October 1, 1984.

Month	DM	CP	AP	ADF	Ca	P	K	Mg	TDN	ME	NEL
	--------------------%--------------------								----Mcal/lb----		
May	22	22	21	28	1.2	.5	1.5	.2	70.4	1.16	.73
June	22	20	19	30	1.0	.4	1.7	.2	67.8	1.11	.70
July	25	21	20	30	1.1	.4	1.7	.3	66.9	1.10	.69
August	24	23	22	28	1.3	.5	1.6	.2	70.4	1.16	.73
Sept.	20	25	23	26	1.4	.6	1.4	.2	71.3	1.17	.74
Avg	23	22	21	28	1.2	.5	1.6	.2	69.4	1.14	.72

Average yield = 3.7 tons DM/acre.

Note: DM = dry matter, CP = crude protein, AP = available protein, ADF = acid detergent fiber, P = phosphorus, K = potassium, Mg = magnesium, TDN = total digestible nutrients, ME = metabolizable energy, NEL = net energy lactation.

Forage energy content would place the pastures of this study in New Zealand's "good" class, and energy supplements should have been needed only for cows producing over 65 pounds of milk per day. But cows producing more than 40 pounds of milk per day received energy supplements, for the same reasons that they were given protein supplements.

Using the forage analyses and the Marshall Ration Balancer feeding tables, which are based on National Research Council standards, we developed feeding programs to incorporate the pasture forage for three farms (Table 10-2).

TABLE 10-2. Feeding programs developed for three Vermont Holstein dairy farms using Voisin controlled grazing management of permanent pastures from May 1 to October 1, 1984.

Feed	Daily milk production/cow, lb						
	30	40	50	60	70	80	90
	--------Daily feed per cow, lb DM--------						
(Farm No. 1: DHIA rolling herd avg = 17,600 lb)							
Hay	6	6	6	6	6	6	6
Corn meal	0	4	9	11	12	14	18
35% Topdress	0	0	2	3	4	5	7
Pasture	27	27	21	23	23	22	20
Total DMI	33	37	38	43	45	47	51
(Farm No. 2: DHIA rolling herd avg = 18,031 lb)							
Hay	4	4	4	4	4	4	
14% Conc.	0	4	9	14	18	24	
Pasture	28	29	26	25	24	25	
Total DMI	32	37	39	43	46	53	
(Farm No. 3: DHIA rolling herd avg = 18,462 lb)							
Hay	3	3	3	3	3	3	3
HMEC	4	6	8	10	13	16	18
Pasture	24	27	29	32	34	34	36
Total DMI	31	36	40	45	50	53	57

DHIA = Dairy Herd Improvement Association estimate of annual milk production. 35% Topdress = 90% DM concentrate containing 35% CP and 80.3% TDN, 1.32 Mcal ME/lb DM, or 0.84 Mcal NEL/lb DM. Pasture = estimate of forage DM consumed, based on what an individual cow would have had to eat to balance her ration, given the amounts of hay and concentrate that were fed. 14% Conc. = 90% DM concentrate containing 14% CP and 78.5% TDN, 1.29 Mcal ME/lb DM, or 0.82 Mcal NEL/lb DM. HMEC = 70% DM high-moisture ear corn ground coarsely. DMI = dry matter intake

Feeding tables used in the United States are influenced by the generally poor forage quality and palatability of continuously grazed pastures. Consequently, the tables greatly underestimate the amount of dry matter that cows will eat in well managed pastures, as well as underestimate the quality (protein and energy) of the forage actually consumed. Because we were working with operating farms, we used existing, accepted feeding tables to develop the feeding programs.

Due to the underestimation of pasture forage quality and dry matter intake by feeding tables, and the farmers' reluctance to further reduce feeding of supplements, our feeding programs were overly conservative, and excess protein and energy were fed. Since they were fed excess supplements, the cows may not have been able to use pasture forage to their fullest potential. In fact, pasture forage intake was estimated on the basis of how much forage the cows had to eat to balance their rations, given the amounts of hay, concentrate, and high-moisture ear corn that were fed in the barns.

Challenging cows to enhance peak milk production usually has been accomplished by feeding more concentrates. If the concentrate ration is reduced or removed, less milk may be produced during the first 3 months of lactation. The farmers in this study were understandably reluctant to reduce the feeding of concentrates to fresh cows very much, because of concern that they might reduce it too much and lower their herd averages. All farmers actually fed less concentrates, however, and some fresh cows did decrease in production, but others maintained their same high production levels.

The fact that supplements were needed at all below production levels where New Zealand criteria would require them, may indicate that selective breeding of cows to use large amounts of concentrates has resulted in animals that are less efficient in using protein and energy from pasture.forage in the United States, compared to cows where such selection has not occurred. Selective breeding is done for milk and butterfat production and type. If these characteristics develop on a high-concentrate diet, it may be possible that breeding in the United States has selected away from efficient forage use. Selecting animals for efficient pasture forage use would seem to be a valid breeding goal, and could result in less protein and energy supplements needed for cows grazing well managed pastures.

Dry matter intakes of pasture forage were adequate for milk production levels up to 80 pounds per day, which included all cows except high-producing fresh ones in the first 3 months of lactation. The dry matter intake needs of high-producing fresh cows could not be met, probably because of the physical limitations of cows in how much forage they can eat in a day. It is nearly impossible for milking animals to eat and process the nutrients required to meet the needs of the udder in producing large amounts of milk. Cows mobilize and use body reserves (milk off their backs) in early lactation to satisfy their energy needs when demand is high. This is a general problem with cows producing more than 80 pounds of milk per day, no matter what the ration is. The high moisture content of the pasture forage also may have limited dry matter intake and made it more difficult for high-producing fresh cows to eat enough dry matter.

With spring-freshening herds, a problem of getting high-producing fresh cows to peak as high on pasture forage as on the more commonly fed silages and hay in confinement, certainly exists. Research is needed to determine how to feed high-producing cows freshening on well managed pasture, to obtain optimum profitability. The important issue is not how high a rolling herd average can be attained at any cost, but rather the profitability of the herd.

257

The other extreme of this feeding problem occurs with herds having low milk production averages. We have observed cows in those kinds of herds peaking at higher milk production levels without any change except placing them under Voisin grazing management, because of the improved quality and quantity of available forage. The problem of dry matter intake by fresh cows is minor for herds with low milk production averages.

Another aspect of this feeding problem is the normal decrease in milk butterfat content that occurs in summer, directly related to lower fiber intake. This problem may become more serious when cows graze lush pasture forage. Milk butterfat content can be prevented from dropping lower than half of what it usually does, by feeding buffers and small amounts of hay or other fiber sources. Farmers in this study fed 4 to 7 pounds of hay per cow per day, divided equally at each milking. One farmer (Farm No. 3) found that by feeding high-moisture ear corn ground coarsely, less hay was needed to maintain a high butterfat content. Coarse grinding allowed ear fiber to be used more efficiently. The ear corn was fed mainly as an energy supplement, but contained enough protein to meet the feeding program requirements as well. Besides hay, 3 to 4 ounces of sodium bicarbonate:magnesium oxide buffer (2:1 mixture) were fed per cow each day.

Since calcium:phosphorus ratios in the pasture forage varied between 2:1 and 3:1 during the season, phosphorus was provided in mineral supplements on all of the farms.

Forage potassium content was adequate, but its magnesium content was low, reflecting the generally low magnesium levels in Vermont soils. Magnesium also was included in the mineral supplements that were fed. The pastures in this study had only been grazed under controlled management for 1 to 4 years and are still improving. Voisin estimated that it would take 100 years for a properly managed permanent pasture to reach its full potential of forage yield and nutritional value. Also, the farmers involved were still learning how to manage pastures properly, and were not grazing intensively enough during much of the season. So forage quality wasn't as high as it might have been.

Three farms (Nos. 1, 2, and 3) had records that enabled us to compare production and economics of controlling grazing to the year before, when they had continuously grazed the same pastureland, besides feeding hay, silage, and green-chopped forage (Table 10-3). Before using Voisin controlled grazing management, farms 1 and 2 had rotated cows through six and four large pasture areas; farm 3 had continuously grazed cows in one large area. All three farmers had discounted most or all feeding value of the pastures, and had fed essentially the same ration throughout the year. Although some cows had been changed on the farms during the elapsed time, certain valid before-and-after comparisons could be made.

When forage from the controlled-grazed pastures was analyzed and the values incorporated into the farms' feeding programs, milk:concentrate ratios of 3.8:1 to 4.4:1 were achieved. Previously the ratios had been 2.6:1 to 2.8:1 throughout the year. Milk produced with such poor ratios is very expensive, with most of the potential profit being lost in buying concentrates.

The increased profitability from controlling grazing was mainly due to feed savings and sales of excess forage. Feed savings included reduced concentrate feeding, and increased stored forage stocks. Feed savings resulted because controlled grazing of the pastures uniformly produced more forage of higher nutritional quality for a longer time, than continuously grazing the same pastures.

Feed costs were calculated for lactating cows only. Calculations were based on the farmers' cost figures, which averaged per ton: $18 for corn silage, $32 for haylage, $70 for hay, and $160 for concentrate.

TABLE 10-3. Production and economics of three Vermont Holstein dairy farms using Voisin grazing management of permanent pastures in 1984, compared to continuously grazing the same pastures in a previous year.

Item	1	2	3	Average
Paddocks				
Average size, acre	3.0	2.0	1.5	2.2
Number needed in spring	13	15	12	13
Number needed by fall	26	30	27	28
Number of rotations	6	5	7	6
Grazing days				
Voisin controlled grazing	129	150	164	148
Continuous grazing	101	87	119	102
Grazing days gained	28	63	45	45
Production factors				
Average number cows in milk	49	24	29	34
Milk/cow/day, lb	50	47	48	48
Milk:concentrate				
Voisin controlled grazing	4:1	4.4:1	3.8:1	4.1:1
Continuous grazing	2.6:1	2.8:1	2.7:1	2.7:1
Butterfat, %	3.4	3.5	3.7	3.5
Average days in milk	174	197	219	197
Income changes				
Feed savings (cows), $	1001	2123	3036	2053
Feed savings (heifers), $	----	1092	----	364
Cash crop (hay), $	2775	----	----	925
Gross/cow, $	77	134	104	105
Gross, $	3776	3215	3036	3342
Expense changes				
Less milk, $	----	799	1914	904
Less butterfat, $	540	----	----	180
(Fencing: amortized, $)	(205)	(545)	(1500)	(750)
Fence maintenance, $	50	50	50	50
Total, $	590	849	1964	1134
Profit: Net/cow, $	65	98	37	67
Net, $	3186	2366	1072	2208

Daily labor, equipment, fuel, and electrical power savings that resulted from using Voisin grazing management weren't included in calculating profitability. Less of these inputs certainly were needed with Voisin management, because less forage had to be fed green-chopped, and less manure had to be cleaned out of barns or from around feed bunks and spread on fields, as compared to feeding in confinement.

Of course, some labor is needed in controlling grazing, but it is nowhere near the amount required for feeding animals in confinement. Calculating labor needs in comparing these situations is somewhat difficult, because people tend not to count the time spend running tractors and equipment as work, but they consider walking out to a pasture to check on plants and move animals as work. This may reflect how sedentary our society has become.

One farmer (No. 3) told me that before switching to controlled grazing from year-round confinement feeding, he would hit the ground running at 4 a.m. and run all day until he crashed at night. One day in mid-July when he was sweating blue blazes, chopping forage, hauling it to a silo, blowing it up into the silo, augering silage out of the bottom of the silo, and hauling it to the feed bunks for the cows, he thought, "Wouldn't it be easier if they did this for themselves?" That was the beginning of the end of year-round confinement feeding on that farm! After switching to pasturing his cows for 5 to 6 months of the year, he said that the tempo of his farm slowed way down, and he was able to enjoy life again.

The amount of equipment and storage facilities needed for farming with Voisin grazing management also is drastically less than what is required for confinement feeding. By changing to grazing his cows from year-round confinement feeding, this same farmer sold about $80,000 worth of corn planting, cultivating, spraying, and harvesting equipment and a silo that was no longer needed! It makes one wonder who gets the best deal when farmers feed year-round in confinement.

Our conservative estimates of net profits in this study ranged from $37 to $98 per cow for the 5-month grazing season. Profits probably would have been even greater if concentrate feeding had been reduced more, if other input savings (e.g. labor, veterinary expenses) had been included in the calculations, and if the cows were capable of using protein and energy from pasture forage more efficiently.

The Atkinson-Thomas Farm

Richard Atkinson and Nancy Thomas have a Jersey farm near East Montpelier, Vermont. In 1983 they began applying Voisin grazing management, using 24 paddocks that average 1.25 acres each. They place milking cows in a fresh paddock each day, and follow them with dry cows and heifers the next day to eat remaining forage. Eight paddocks are set aside for machine harvesting of surplus forage in mid-June. In the future, they intend to give cows a fresh paddock after every milking. This study is based on their experience with 38 cows and 10 heifers for a 183-day grazing season between May 5 and November 4, 1988.

This economic analysis uses a "without versus with" method (Tables 10-4 to 10-6). "Without" refers to continuous grazing with little management. "With" occurred after 4 years of using Voisin grazing management.

Richard and Nancy found that using Voisin grazing management significantly increased productivity of their pasture, by extending their grazing season 35 days. This resulted in a savings of 18 tons of hay that otherwise would have been fed in the barn. Because of improved pasture forage quality, less protein concentrate had to be fed, resulting in a savings of $20/ton of concentrate.

A lot of labor was saved by controlling grazing, from

feeding less hay and handling less manure. Of course, less fuel was needed and less wear of machinery occurred.

The greatest benefit realized was in the amount and quality of milk that their cows produced. Milk production increased an average of 1260 pounds per cow during the grazing season. Protein in the milk increased 1.2 percent. These improvements amounted to a net profit increase of $4648 per year, directly attributable to using Voisin grazing management.

The total annual net profit increase from the above benefits was $7866.

Other substantial benefits occurred, but a dollar value wasn't calculated for them, and they weren't included in the economic analysis. Richard and Nancy feel that their animals are in better condition under Voisin grazing management. They can monitor their animals better for heat detection, and the animals have fewer health problems. This saves money on breeding and veterinary expenses. All animals (especially heifers) handle better because they're worked with daily under controlled grazing. The ease of operation for the farm generally improves during the grazing season, because Richard has more time to manage the cows.

Richard and Nancy feel that an environmental benefit also is derived from using Voisin grazing management. This is because a forage source with little or no soil erosion (pasture) replaces sources associated with a great deal of soil erosion (corn and other annual row crops). They also feel that water quality benefits, because nutrients are held and used better in a pasture than in row crops such as corn.

Richard and Nancy have found Voisin controlled grazing management to be a low-cost input alternative to high-cost, capital-intensive dairy farming. Although a capital-intensive operation with year-round confinement feeding may increase milk production, higher input costs for machinery and labor could negate any additional income. The lower costs associated with Voisin grazing management make it a profitable and sustainable way of farming over the long term. (Pillsbury and Burns)

TABLE 10-4. Economics of using Voisin controlled grazing management on the Atkinson-Thomas farm in Vermont.

Benefits from using Voisin grazing management	$7866
Costs of using Voisin grazing management	$2093
Net benefit	$5773
Benefit: cost ratio	3.8:1
Net benefit per cow	$ 32
Net benefit per grazing day	$ 152

TABLE 10-5. Annual benefits of using Voisin grazing management on the Atkinson-Thomas farm in Vermont.

Reduced feed costs:	Replaced hay & silage	$1715
	Savings in concentrates	278
Reduced labor in feeding†		350
Reduced manure handling††		840
Reduced bedding		32
Increased milk production		4648
Total benefits		$7866

†Labor saved in unloading and feeding one truckload of hay (800 bales), valued at $5/hour.
††Includes labor saving of $240 ($5/hour) and machinery expense of $600, including everything from running gutter cleaner to spreading manure in the field.

TABLE 10-6. Costs of using Voisin controlled grazing management on the Atkinson-Thomas farm in Vermont.

Extra animal handling (moving animals)	$ 375
Mowing and harrowing paddocks (1-2 times/year)	390
Fertilizer (192 lb 0-10-40/2 years)	385
Lime (2 ton/acre/5 years)	300
Fence (amortized energizer & portable fencing; perimeter fence already existed)†	111
Additional hay for dry cows††	334
Additional concentrate for dry cows††	98
Gravel for lanes (14 cubic yards)	100
Drinking water (no added cost; same as before)	
Soil tests (no cost to farm, but done every 2 years)	
Total costs	$2093

†If perimeter fence had been needed, it would have cost $1600; amortized over 30 years this would be $170/year.
††Average of 25 dry cows for 3 weeks before calving; hay: 17 lb/day @ $75/ton, including labor to feed; concentrate: 2.5 lb/day @ 150/ton, including labor to feed.

The Opitz Farm

Charlie Opitz feeds 800 Holstein milking cows and 1200 dry cows and heifers on 2000 acres near Mineral Point, Wisconsin. He produces 12- to 13-million pounds of milk per year. During 7 to 8 months of the year the cows and heifers get most of their feed by grazing under controlled management. In 1989 Charlie calculated that his farm produced about $800 worth of milk per acre. Some of the better pastureland returned up to $1300 per acre to management and labor, after costs for seed, fertilizer, purchased feed were deducted!

The decision to pasture the milking cows came in 1987,

when Charlie was dealing with heat-stress problems in the confinement barn. He put 80 milkers on pasture under controlled grazing management in the spring of that year just to see what would happen. During one hot spell, milk production from cows in the barn dropped 22 percent, while production from the grazing cows fell only 12 percent. The 80 grazing cows made $900 worth of milk per acre above supplemental feed costs in 6 months grazing! That was enough for Charlie; now all cows graze.

Grazing on this farm starts when the grass is 2 inches tall (about 1400 lb DM pasture mass). In 1989 heifers started grazing in late March, and didn't come off pasture until December 21. Milking cows start grazing on about April 24, and stop grazing sometime in November, depending on the season. Starting the first rotation early is essential to get staggered regrowth for the whole season. Milking cows graze a paddock first for 1 to 2 days, followed by dry cows and heifers for 3 to 6 days. Recovery periods between grazings vary, depending on plant regrowth and pasture mass (6 inches tall). Heifers are also used to clean up hay fields right after the hay is removed, thereby gaining a day or two in recovery time for the pasture.

Charlie prepares for spring grazing beginning in the previous autumn. He stops grazing about two-thirds of his pastures in early September. Half of the deferred pasture is grazed later in that fall, and the other half is carried over winter to be grazed early in the following spring. In the spring, forage on the deferred pasture still contains 14 to 18 percent crude protein! Deferring pasture from fall to spring enables Charlie to begin grazing early on well drained soil.

The rolling herd average milk production is 14,000 to 15,000 pounds. Before putting the cows on pasture, the herd average was 17,000 pounds when cows were milked three times a day. Feed and other cost savings have more than made up for the decrease in milk production. Now that cows are being milked twice a day, only six or seven people are needed to milk, compared to the 12 needed before.

During the grazing season, half of the cows' feed comes from pasture, and half is fed in the barn at milking. The ration for milkers consists of 6 to 7 pounds of alfalfa hay, 1 to 2 pounds of sudax or small grain (wheat) silage, and 12 pounds of concentrate containing varying combinations of wheat middlings, hominy, wet gluten, distiller's grain, and full-fat soybeans. In winter the concentrate portion is increased to 24 to 27 pounds.

Charlie double-crops the sudax and wheat, to produce about 40 tons of silage per acre. Although hay yields have been 4 to 5 ton DM/acre, a 1990 estimate of pasture forage production indicated that the yield may be twice as high under grazing! Analyses showed that the pasture forage averaged 22% dry matter, 22% crude protein, 24% acid detergent fiber, and 40% neutral detergent fiber during the 7-month grazing season.

Charlie estimated that it cost $40 to $70 per acre to install fencing and drinking water. He says that's a bargain, considering that it costs about $20 a trip to cut silage or $100 per acre to make hay each year. He feels that farmers can pay off their land debt, just with the savings that they can make in operating costs from using controlled grazing management. Annual fuel consump-tion on his farm is now $30/cow, compared to the Wisconsin state average of $60/cow. He used to cut about 3000 acres (including rented land) of hayland each year. Using controlled grazing management has reduced the need for hay so much that now he only cuts 1200 acres of hayland. Following his estimate of how much it costs to make hay per acre, Charlie makes a gross saving of $180,000 a year (3000 - 1200 = 1800 acres x $100) on reduced haying costs. No wonder he smiles a lot!

As a result of his grazing management, Charlie's

267

pastures now contain mostly quackgrass, bromegrass, tall fescue, orchardgrass, bluegrass, alfalfa, red clover, and birdsfoot trefoil. He feels that there is no optimum grass or legume for the entire farm. A combination of plant species is needed to cover the shortcomings of each.

He has come to prefer bromegrass in his pasture. If it has nitrogen from associated legumes or manure, bromegrass contains up to 21 percent crude protein, and can provide five grazings a year, carrying nearly 2 cows per acre. Quackgrass also performs very well for Charlie. He grazes it five times on high fertility soil, and it often contains more than 30 percent crude protein!

More fun and profit aren't the only benefits that Charlie has realized since he went to Voisin controlled grazing management. Local Soil Conservation Service officials estimate that Charlie's land was losing about 90 tons of soil per acre per year before he took the slopes out of corn production and seeded them to perennial forages. This not only stopped erosion and silting, but also almost entirely eliminated pesticide applications. Charlie says that grass farming solves 99 percent of the problems that the USDA's Low Input Sustainable Agriculture program is trying to deal with. Once the source of problems (annual row cropping, esp. corn) is removed, symptoms disappear.

Charlie Opitz's farm is proof that Voisin controlled grazing management isn't just for small herds, but is a profitable alternative for any size dairy farm. (Cramer, Liebhardt)

Farm Management Worksheets

John Cockrell (the Extension Agent who works with Charlie Opitz) has developed the following worksheets of farm management alternatives for an average Wisconsin dairy farm, that should be helpful in your planning (Tables 10-7 to 9). It's clear that feeding milking cows on pasture as much as possible is the most profitable farm management strategy. In the examples given, the net profit for year-round confinement feeding is only $141 per cow. When forage and grain are purchased and some pasture is used, profit increases to $688 per cow. When pasture is used to provide most of the feed for the cows, profit will be at least $937 per cow!

Actually, the example in alternative II of 50 cows on 150 acres of pasture, greatly underestimates the productive potential of the pasture, and the profitability of this alternative. With Voisin controlled grazing management, stocking rate can be at least 1 cow/acre, so only about 50 acres of pastureland are needed for 50 cows. This also would include harvesting surplus forage from about half (25 acres) of the grazed land in the spring. So surplus forage could be (would need to be) harvested on about 125 acres of this farm. If we estimate a conservative hay yield of 4 ton/acre on 100 acres and 2 ton/acre on 25 acres, that comes to 450 tons of hay. If the 50 cows and 10 replacements eat 50 lb hay/day during 150 days of winter feeding, there would be about 225 tons of hay that could be sold. If that were sold for $75/ton, an additional gross profit of $16875 could be realized! (150 days @ 50 lb = 7500 lb/2000 = 3.75 ton x 60 animals = 225 ton; 450 - 225 = 225 ton; 225 x $75 = $16875)

Alternative I of 50 cows on 50 acres of pasture, plus purchased forage and grain, similarly underestimates the productive potential of the pasture. Much less forage and grain would need to be purchased than was projected, so profit per cow would be greater.

WORKSHEET 10-1. Estimated annual cost to produce corn, alfalfa, and small grains on a Wisconsin dairy farm that has 200 total acres, 150 tillable acres, and carries 50 to 60 cows plus replacements in year-round confinement.

	Example†	Your farm
Machinery: ($100,000 - $175,000)		
average = $137,000, x 10% interest	$13750	$_____
Depreciation: 7 years	19642	_____
Repairs (10% of value)	13750	_____
Seed, chemicals, fertilizer ($75/acre)	11250	_____
Crop insurance ($7/acre)	1050	_____
Fuel and electricity	4000	_____
Taxes ($30/acre, 150 tillable acres)	4500	_____
Land cost ($800/acre, 10% interest)	12000	_____
Total costs	$79942	$_____
Average production/cow:		
14500 lb milk @ $12/cwt =	$ 1740	$_____
Average cost/cow: $79942/50 cows	1598	$_____
Net profit/cow	$ 141	$_____

†Labor and storage costs are not included.

WORKSHEET 10-2. Alternative farm management I: estimated cost to purchase forage and grain on a Wisconsin dairy farm that has 50 cows and replacements, and uses 50 acres of pasture.

	Example	Your farm
Machinery (tractor & spreader)		
15000 x 10% interest	$ 1500	$_____
Depreciation: 7 years	2142	_____
Repairs (10% of value)	1500	_____
Fuel and electricity	500	_____
Purchased forage (9 tons/cow)	32000	_____
Grain (6000 lb @ $0.05)	15000	_____
Total costs	$52642	
Average production/cow:		
14500 lb milk @ $12/cwt	$ 1740	$_____
Average cost/cow: $52642/50 cows	1052	_____
Net profit/cow	$ 688	_____

WORKSHEET 10-3. Alternative farm management II: estimated cost to pasture 50 cows and replacements on a Wisconsin dairy farm using 150 acres of pasture.

	Example	Your farm
Machinery (tractor & spreader)		
15000 x 10% interest	$ 1500	$_____
Depreciation: 7 years	2142	_____
Repairs (10% of value)	1500	_____
Seed and fertilizer	3000	_____
Fuel and electricity	500	_____
Taxes ($30/acre x 150 tillable acres	4500	_____
Land cost ($800/acre,10% interest)	12000	_____
Grain (6000 lb @ $0.05)	15000	_____
Total costs	$40142	$_____
Average production/cow:		
14500 lb milk @ $12/cwt =	$ 1740	$_____
Average cost/cow: $40142/50 cows	803	_____
Net profit/cow	$ 937	_____

Note: The farm has enough pasture forage for at least 100 cows; it could be very profitable if the herd size was increased and replacements were purchased.

The Adamski Farm

Rick Adamski feeds 40 cows and 20 replacement heifers on pasture under controlled grazing management on his 200-acre farm near Seymour, Wisconsin. He is convinced that feeding cows on well managed pasture is more profitable than feeding them in confinement.

In 1988 (a very dry year) Rick's cows grazed for 161 days. During that time they produced 2762 cwt of milk, and consumed 47860 lb of grain. This averaged out to 17.3 lb grain/cwt milk, or a milk:grain ratio of 5.8:1. The average amount of grain fed in Wisconsin was 37 lb/cwt milk, or a milk:grain ratio of 2.7:1. Compared to this average, Rick estimated that he saved 54000 lb of grain, or $3780, based on a grain cost of $7/100 lb.

Since fuel and electricity can be saved by feeding livestock on pasture rather than in confinement, Rick calculated fuel savings in terms of gallons of gasoline and diesel fuel. Feeding cows on pasture, instead of hauling hay to a barn or silo and then feeding it, saved 105 gallons of gasoline and 330 gallons of diesel fuel. At $1/gal of each fuel, Rick estimated that he saved $435.

Analyses showed that the forage in Rick's pasture averaged 26% dry matter, 20% crude protein, 28% acid detergent fiber, 42% neutral detergent fiber, and 67% total digestible nutrients during the grazing season. The high quality of pasture forage was reflected directly in low feed costs, which ranged between $1.97 to $2.58/cwt of milk produced during the grazing season. (Schieldt)

SEASONAL DAIRYING

Ohio State University

Dr. David Zartman, Head of Ohio State University's Dairy Science Department will complete a 5-year study on seasonal dairying (e.g. spring-calving and producing most milk on pasture) in 1991. His findings already show that by feeding 60 spring-calving cows on pasture, a farmer would net $36000 per year, plus have a vacation from milking for 2 months during the winter when the entire herd is dry!

Milking cows are allowed just enough pasture forage so that they consume about 60 percent of the available forage in a 24-hour grazing period. Then they are given a fresh paddock, and heifers follow behind grazing the remaining forage. Recovery periods range from about 15 to 30 days, depending on plant regrowth. In the spring, surplus forage is harvested as hay for winter feeding.

The key to seasonal dairying is synchronizing milk production with pasture forage availability. The cows should be at their greatest nutritive requirement when pasture plants are growing the fastest, and the forage has the highest protein and energy levels. According to Dr. Zartman, farmers have been incorrectly sold the monoculture farm management model that includes feeding, breeding, and managing cows in the same way to get maximum milk production out of them no matter where they are. All evidence indicates that the model is invalid. He says that farmers need to try alternatives that fit local environments and economic needs.

At the experiment farm near Youngstown, cows receive 1 pound of 14% protein supplement per 5 pounds of milk produced, or a minimum of 10 pounds of supplement per day. Dr. Zartman keeps supplemental energy levels a little short, so that the rumen micro-organisms do a better job digesting forage. He says that production is somewhat less, but it's more efficient economically.

In the first year of the study, commercial grade Holsteins averaged 15154 pounds of milk, and Jerseys averaged 11353 pounds. During the second year, pasture quality had improved, and milk production increased to 17865 and 12458 pounds.

Cows and heifers are synchronized for breeding with Synchromate-B implants. They are bred to begin calving on March 10. Any animals that breed late and aren't due to freshen by May 10 are sold. The herd is dried off on December 20, after an average lactation of 270 days.

One problem with seasonal dairying is that all of the calves arrive within a short period. But at that time there are no crops to worry about, and once grazing begins barn chore demands decrease greatly. The other problem is that the price of milk is lowest when most of the milk is shipped. But that's acceptable because the milk is produced at a much lower cost on pasture than it is in confinement.

Benefits include improved herd health, with little or no mastitis. Cows aren't subjected to the stress of calving in winter, and frozen teats are no longer a problem. The best part of seasonal dairying on pasture is its low capital requirement. All that's needed is a barn, well managed pasture, a manure spreader, and a tractor. Supplements, hay, and silage can be purchased, or forage can be custom-harvested for storage as hay or silage.

Dr. Zartman has developed a computer spreadsheet that farmers can use to analyze their records and project how changes in management or pricing will affect their profitability. For more information write to Dr. Zartman at the Dairy Science Department, Ohio State University, Columbus, OH 43210-1094. (Cramer)

The Patenaude Farm

Dan and Jeanne Patenaude farm in the hills of southwestern Wisconsin. Nestled into the base of a forested ridge, their farm lacks many of the things generally associated with conventional dairy farms. There are no silos and no manure storage pit, no large shed filled with big tractors and machinery, no plowed land for growing corn or grain, and no debt. Instead there is a small well maintained barn and a small shed for storing the tractor, haybine, and baler.

The other things that make this farm different from conventional dairy farms is its low labor requirement, high profitability, and synchronized breeding for spring calving to produce most of the milk on pasture.

The farm has 73 acres, with only 27 of those being tillable. There are 20 acres of permanent pasture, and the rest is woodland. A trout creek flows through fertile bottomland that used to grow corn, but now is covered with lush grass-red and white clover pasture.

The 24 high-producing (17000 lb average) Holsteins and young stock grazed for 223 days (April 13-Nov. 21) in 1988 and 215 days in 1989. Dan gives the cows a fresh paddock after one or two milkings. Dry cows and heifers follow the cows to clean up remaining forage. Analyses showed that the pasture forage contained 20 to 26% crude protein and 0.70 Mcal NEL/lb DM.

Dan estimated feed costs to be $2.76/cwt of milk during the grazing season, calculated from a grain cost of $1.67/cwt and a forage cost of $1.09/cwt. This is about $4/cwt of milk lower than winter feed costs, and directly reflects the use of inexpensive pasture forage.

Because there were no satisfactory fencing materials available in Wisconsin when Dan and Jeanne fenced their paddocks, Dan became a dealer of New Zealand style fencing equipment. Since it takes so little of his time to care for his cows during the grazing season, Dan now has time to help other farmers design pasture layouts and set up fencing.

If it costs so much less to feed cows during the grazing season, why not produce most of the milk during that time, instead of working like a maniac to do it during the winter? When Dan asked himself that question, the next step was obvious. In autumn 1990 he sold all of his fall-calving cows, keeping or buying replacements that would begin calving in late February. He dried the whole herd off just before Christmas and took a well deserved and needed 2-month vacation from milking. (McNamara, Liebhardt)

The Moore Farm

Austin and Linda Moore have farmed near North Whitefield, Maine since 1970. The Moores have always kept debt and the need for hired labor to a minimum. They have never grown corn. Those things alone would place Austin and Linda in a unique group among dairy farmers, but they have never stopped being innovative in their farming practices. Their latest innovation occurred in 1988 when they sold their herd of registered fall-calving Holsteins (18000 lb average), and bought spring-calving Jersey heifers. They did this to take advantage of the cheap feed available during the grazing season to produce most of their milk, the greater efficiency of Jerseys, and to be able to take a vacation in winter when the herd is dry.

The farm contains 180 acres, with 100 acres of hayland, 28 acre of permanent pasture, and the rest in woodland. During the first 10 years on the farm the Moores grazed 30 to 33 Holsteins continuously in three to four open pastures. With this method, cows began grazing in mid- to late-May, and by August had to be fed supplemental hay because the pasture plants stopped growing. Then in mid-summer 1981 Austin heard about controlled grazing management at a meeting. The next day he bought an energizer and began subdividing the pastures into paddocks. By 1982 they had converted fully to controlled grazing, with 21, 1-acre paddocks, each equipped with drinking water. The cows occupy paddocks for 24 hours, and graze from late April until about October 15.

Grain concentrate feeding on the farm has declined steadily since the Moores began applying Voisin controlled grazing management. In 1986 the average annual consumption of concentrate was 48% less than what it had been during 1978 to 1980, when the milk:grain ratio was 2.4:1. Between 1984 through 1986, the ratio had risen to 3.3:1, which is an increase of 83% more milk per pound of grain fed. Consequently, the cost of producing milk is much lower when cows are fed on well managed pasture. (Gage and Smith, Merrill)

277

BEEF

Cow/Calf Extended Grazing

Extended grazing usually involves grazing hay fields (meadows) during times of the year when hay otherwise would be fed to a herd. The longer that grazing lasts during a season, the less forage must be mechanically harvested and stored for winter feeding. Making maximum use of pasture by extending the grazing season reduces production costs and increases profitability. This 3-year study was done by West Virginia University researchers to determine which meadow management practice gives the most profit.

Four management practices were evaluated: 1) two cuttings of hay, no grazing; 2) one hay cutting, late fall grazing; 3) early spring grazing, two hay cuttings; and 4) early spring grazing, one hay cutting, and late fall grazing. The practices were compared for orchardgrass and tall fescue.

Total forage production for both orchardgrass and tall fescue (hay + grazed) was highest (3.4 ton DM/acre) when meadows were cut once and grazed in late fall. Forage production was lowest (2.9 ton DM/acre) when meadows were grazed in early spring and cut twice. Tall fescue consistently produced more forage for haying and grazing than orchardgrass.

With the exception of costs on a per ton of hay basis, where the costs of cutting once and grazing in fall were lowest, the least-cost practice was spring grazing, one cutting, and late fall grazing. Spring grazing followed by two cuttings had the highest costs on a per ton of hay basis. But two hay cuttings with no grazing was the highest cost method, regarding all other aspects, such as costs per acre, per ton of dry matter, and per pound of calf produced. Except for costs on a per acre basis, average production costs were lower for farms using tall fescue than those with orchardgrass. Production costs were lower when hay was baled in 1500-lb round bales, than when baled in 50-lb square bales.

The practice of haying and grazing once resulted in the most calves sold. This is because more forage was produced with this method. The most profit in terms of returns to fixed resources, however, was obtained when meadows were grazed in spring, cut once, and then grazed in fall, for both orchardgrass and tall fescue. The least profit resulted when meadows were hayed twice and not grazed (tall fescue), or grazed in spring and hayed twice (orchardgrass).

The researchers concluded that early spring grazing, one hay cutting, and fall grazing is the most profitable way of extending the grazing season for cow/calf production. They also found that tall fescue generally performed better than orchardgrass in terms of forage production, cost, and profitability. (D'Souza)

SHEEP

It's difficult to make a profit with a small flock of sheep. But sheep don't have to be raised for profit. Horses, dogs, and cats usually aren't kept for profit, and few people raising small numbers of pigs, goats, or beef cattle make an actual profit. Raising animals can be rewarding and beneficial in aspects other than calculations of profit and loss. If we would assign economic value to such things as enjoyment and keeping land clear, it would be easy to show a profit for raising sheep. But many people raise sheep expecting to make an actual dollar profit, only to [...] ernatives to [...] commercial [...] all flock of

SHEEP ECONOMICS: I didn't include a discussion of Peter & Hilary Wood's farm in Wisconsin. They achieve a net return of $82.20/ewe/yr with 250 Finn x Rambouillet ewes, from a 300% lamb crop and pasture feeding April 15- Jan. 15! For their story see *The Shepherd* March 1990 & Jan. 1991.

sheep, milk them. Sell liquid milk or process the milk into cheese, yogurt, or fudge and sell the products. You'll make money.)

There are many aspects to achieving a profit with sheep, and many people have failed to do so, even though they tried hard. Since providing feed is the largest single cost in raising sheep (and all other livestock), it's the aspect where most can be gained by improving the feed supply. Whatever your reason for raising sheep, the costs involved can be reduced greatly by feeding them on well managed pasture as much as possible. (Mitchell)

The Blazek Farm

Michael and Terri Blazek raise a small flock of Suffolks on a farm near Waldoboro, Maine. Their main objective with the sheep is to produce high value breeding stock for sale. They converted to Voisin controlled grazing management in 1985, after raising sheep for 8 years with conventional practices. Before they began controlling grazing they mainly used the land for producing hay, with some aftermath grazing.

This economic analysis was based on a "before" versus "after" approach. In each case the Blazeks used the same 10 acres of land. Before they began controlling grazing, very little forage was obtained from the land. Eleven ewes weighing 200 pounds each grazed for about 45 days per season. About 320 bales of poor quality hay was also harvested from the 10 acres. Additional needed feed was purchased.

In 1985 the Blazeks seeded 1 acre to Grimalda perennial ryegrass, and began controlling the grazing of their sheep late that spring. They built a lane down the middle of their land, dividing it into two 5-acre areas. They placed black plastic tubing with four spigots in the lane to provide drinking water with a portable tub to paddocks. They formed twelve, 3/4-acre paddocks with portable fencing that they could remove to make hay if necessary. Three acres of land continued to be used mainly to produce hay.

Controlled grazing management improved the quality and quantity of forage produced on the farm. Since most of the land was used for pasture, the amount of hay harvested declined to 160 bales. But now 22 ewes (200-lb each) graze for 200 days, 9 lambs (135-lb each) graze for 45 days, and 2 rams (300-lb each) graze for 107 days. This totals 999 animal unit days.

(An animal unit day is the amount of feed needed to support 1000 pounds of animal weight for one day. For example, one animal unit day is the amount of feed needed per day for 8 ewes weighing 125 pounds each.)

To estimate the value of increased production, the additional animal unit days were valued in terms of their hay equivalent. The total cost of feed (pasture, hay, grain) needed before and after applying controlled grazing was determined and divided by the number of animal units supported in each situation. The cost of feeding one animal unit decreased from $338 without controlled grazing, to $170 with controlled grazing. With 200-lb ewes, this meant that annual feed cost was reduced 50 percent, from $68 down to $34 per ewe. This cost savings was multiplied by the number of animal units that the Blazeks now can support, to obtain the increase in feed value of their land under Voisin controlled grazing. The increased value amounted to $1042 per year.

Because the sheep were outside on pasture more under controlled grazing, time needed for manual manure handling decreased by 51 hours. Valued at $5 per hour, this saved $255. For the same reasons, bedding expense decreased by $200, and time needed to feed hay and grain was reduced by 100 hours, for a saving of $500.

Costs of converting to controlled grazing included

fencing, the water supply, seeding one acre, moving animals every 4 days, and fertilizing the area where hay was cut.

Voisin controlled grazing management proved to be very beneficial to the Blazeks. The economic advantages will continue to improve, because they intend to increase the number of animals carried on the 10 acres. So this analysis is conservative (Tables 10-10 & 10-11).

Although not quantifiable, benefits were gained in other areas besides feed and labor savings. Michael and Terri feel that Voisin grazing enables their sheep to stay in better condition, and especially helps the ewes to recover faster from lambing. They feel that the quality of the pasture has improved drastically, with much higher white clover content, and the quality continues to improve. Since they handle and observe the animals more in moving them among paddocks, they feel that their livestock management has improved. This is especially valuable to them, because of their breeding stock operation. (Burns and Jones)

TABLE 10-7. Economics of converting to Voisin controlled grazing management from conventional practices on the Blazek sheep farm.

1. Additional costs of controlled grazing
   A. Fence (amortized to annual equivalents)

Permanent	$290
Portable	40
B. Seed (1 acre)	58
C. Water system (amortized)	27
D. Moving animals (includes providing water)	65
E. Fertilizer (machinery borrowed, no cost)	404
Total annual costs	884
Lower haying costs	-144
	$740

2. Benefits received from controlled grazing

A. Improved feed value of 10 acres	$1042
B. Reduced manure handling	255
C. Reduced bedding expense	200
D. Reduced labor in feeding	500
Total annual benefits	$1997
Less hay produced	- 80
	$1917

3. Comparison of costs versus benefits of controlled grazing

Annual benefits	$1917
Annual costs	$ 740
Net annual benefits	$1177

The Blazeks estimated that they received about $2.60 in benefits for each $1.00 invested in controlled grazing ($1917 ÷ 740 = 2.6).

TABLE 10-8. Comparison of feed per animal unit (AU) by source of feed on the Blazek farm before and after converting to Voisin controlled grazing management.

Pasture
    Before $54 ÷ 3.31 AU = $16/AU
    After $549 ÷ 6.22 AU = $88/AU
Hay
    Before $825 ÷ 3.31 AU = $249/AU
    After $330 ÷ 6.22 AU = $53/AU
Grain
    Before $240 ÷ 3.31 AU = $73/AU
    After $180 ÷ 6.22 AU = $29/AU

Summary:

Feed source	Cost per animal unit, $		
	Before	After	% change
Pasture	16	88	+450
Hay	249	53	-79
Grain	73	29	-60
Total	$338	$170	-50

CONCLUSION

The evidence is overwhelming: Voisin controlled grazing management works and it pays! Don't accept pastures for what they are any longer. Think of what they could be if you would manage them with the same amount of attention that you give to your other cropland.

REFERENCES

Burns, P. and C.R. Jones. 1987. Voisin rational grazing - sheep (Maine economic case study). Soil Conservation Service, Orono, Maine. Mimeo. 44 p.

Burch, C. 1984. Dairy sheep update. Sheep!. 5(3):18-19.

Cockrell, J.R. 1990. Worksheets of farm management alternatives for an average Wisconsin dairy farm. University of Wisconsin Extension Service, Darlington, Wisconsin. Mimeo. 3 p.

Cramer, C. 1990. Grass farming beats corn! -- The Charles Opitz Wisconsin dairy farm. The New Farm. Sept./Oct. p. 10 - 16.

Cramer, C. 1990. Milk 10 months, take a vacation...net $36000! (Ohio State University Seasonal Dairying) The New Farm. May/June. p. 12-14.

D'Souza, G.E., E.W. Maxwell, W.B. Bryan, and E.C. Prigge. 1990. Economic impacts of extended grazing systems. American Journal Alternative Agriculture. 5(3): 120-125.

Gage, S., and S. Smith. 1989. The Moore dairy farm. American J. Alternative Agriculture. 4(1): 35-37.

Jordan, R.M., and W.J. Boylan. 1986. Sheep milk, cheese and yogurt production. Proceedings, "Adapt 100", Des Moines, Iowa. Successful Farming Magazine. December.

Liebhardt, B. 1991. Rotational grazing - Case studies. In press.

Merrill, L.S. 1990. The thinking farmer: doing things that others say can't be done. (The Moore Dairy Farm). New England Farmer. August. p. 6-8.

Mitchell, D. 1982. *So You Want To Raise Sheep In Vermont*....Vermont Sheep Breeder's Association. Handbook. 48 p.

Murphy, W.M., J.R. Rice, and D.T. Dugdale. 1986. Dairy farm feeding and income effects of using Voisin grazing management of permanent pastures. American Journal of Alternative Agriculture. 6(1): 147-151.

Pillsbury, B.P., and P.J. Jones. 1989. Economics of adopting Voisin grazing management on a Vermont dairy farm. USDA Soil Conservation Service, Winooski, Vermont. Bulletin. 9 p.

Schieldt, L. 1989. Back to pasture - the Rick Adamski farm. Wisconsin Agriculturist. January. p. 16-17

RENOVATION

The flowers depart
When we hate to lose them;
The weeds arrive
While we hate to watch them grow.

Dogen

Whenever anyone mentions pasture improvement in the United States, almost everyone thinks of renovation. The renovation process includes at least partially destroying the sod, plus fertilizing, liming, seeding into a prepared seedbed or into the killed sod with a sod-seeder, and controlling weeds. In short, doing whatever is needed to establish desirable plant species. Destroying the sod involves either 1) severe overgrazing to weaken the sod, followed by several passes with a heavy disc at different times of the year to cut the sod into pieces, or 2) applying herbicides that kill all or most of the plants, including desirable species.

In renovation instructions, some suggestion may be made to change the grazing management so the pasture doesn't revert to the same mess that it was before. But the suggestions generally have been vague and involved using a few large paddocks, long occupation periods, and fixed recovery periods. This kind of management doesn't take into account the needs of plants, and consequently fails to maintain renovated pastures in an improved condition.

Is renovation as conventionally attempted, practical or even necessary?

Conventional renovation practices essentially are attempts at transferring field-cropping techniques to a pasture situation. They ignore the fact that a pasture is a different ecological environment, compared to that of monoculture field crops grown on plowed and cultivated soil. Destroying a pasture sod and/or plowing and cultivating soils, drastically changes conditions in soils and disrupts balanced relationships among the organisms that live there and make soils alive. It takes a long time after plowing a pasture sod for the relationships among soil organisms to reach the equilibrium, and for the soils to return to the same conditions, that existed before plowing. (Voisin wrote that it would take about 100 years, if the pasture is well managed after renovation!)

Andre Voisin may have been the first person to question the practice of pasture renovation, when he wrote that plowing up a pasture does not make up for defective grazing management. Voisin stated that the logical approach to improving pastures is to change the grazing management, and wait a couple of years, to see how the pasture plant community responds. He cited observations by other ecologists and results of several research studies, which clearly showed that pasture plant communities are extremely dynamic and change very rapidly with the grazing management applied.

One study cited by Voisin is especially interesting because it was done at Cornell University, and should have influenced thinking about pasture improvement in the United States. The Cornell study involved pasture plots all sown with the same simple mixture of Kentucky bluegrass and white clover. Plots were cut to 1/2 inch above the soil surface at different fixed intervals. After two years of cutting treatments, clover contents in the plots ranged from 1 to 80 percent, depending on the cutting management used. This was one study among many others done elsewhere, which showed that the botanical composition of pasture swards depends mainly on recovery periods between cutting or grazing.

There's good evidence that existing pasture plants can be very productive if managed well. For example, recent research in Scotland and England showed that unless a farmer is willing to apply more than 110 pounds of nitrogen/acre/year, renovation to introduce new cultivars of perennial ryegrass can't be justified. If managed well, grasses (e.g. bromegrass, timothy, orchardgrass, Kentucky bluegrass, and quackgrass) already existing in permanent pastures produce just as much forage and protein per acre as new selections of perennial ryegrass without heavy nitrogen fertilization.

Besides, I have yet to see anyone in the United States fully utilizing the increased amount of forage that's produced when Voisin controlled grazing management is applied. There's no sense in spending money, time, and labor to renovate a pasture to promote higher production, if you don't completely use the forage that's already being produced.

Another aspect ignored when pasture renovation is advocated, is that most permanent pastureland in the northern United States is so-called marginal land. The reason it's considered to be marginal (and is in permanent pasture) is because it has steep, rocky, or shallow soils, ledge outcroppings, boulders, brush, or trees that limit or prevent its use for cultivated field crops. These problems make the conventional field-cropping techniques of renovation very difficult or impossible to use on most permanent pastureland that needs improving.

One of my attempts at improving pastures involved frost seeding white clover in an old meadow, without seedbed preparation. (Frost seeding takes advantage of the freezing and thawing action that opens and closes soil pores in the early spring. The soil movement covers seed that simply was broadcast on the frozen soil just before or right after the snow melts.) The meadow was then grazed with sheep under controlled management. By October, clover content in the plant community had increased from its original 2.5 percent, to about 25 percent where clover had been seeded.

But to get to the frost-seeding experiment, I had to walk across pasture areas that had been grazed under Voisin management for 3 years with cattle. I noticed that in those areas white clover content had increased to about 30 percent, without any kind of seeding or any other treatment except changing the grazing management! By the end of the second year of the frost-seeding experiment, white clover content in all plots had increased to about 30 percent, regardless if clover had been seeded or not.

This experience with pasture improvement, and several years of observing the desirable changes in plant composition and increases in forage yield that generally occur in Vermont pastures when Voisin grazing management is used, have persuaded me that the traditional method of attempting to improve pastures (i.e. renovate and then maybe change the grazing manage-ment) is just the opposite of what it should be.

There are, of course, situations where little legume or desirable grass seed might exist in the soil, such as in long abandoned pasture, old tall grass hayfields, or tilled cropland where herbicides have been applied. On tilled cropland you can seed whatever grasses and legumes you want into a prepared seedbed. Where a sod already exists, you can speed the transition to a more productive sward by frost seeding the legumes and grasses that you want to introduce. This method won't work though, if the soil surface is covered with a mat of undecomposed plant residue that prevents good soil and seed contact. In this case, you simply need to change your grazing management for at least one season until the residue mat disappears, before attempting to seed anything.

Frost-seeding success also varies with soil type, plant species, and competition from existing pasture plants. Heavy, moist soils heave more and may cover seed better than coarse textured, drier soils. Legumes are easier to frost-seed than grasses. It is very difficult to introduce new grass cultivars into an existing, vigorous pasture sward. For example, I've introduced red and white clover into my existing swards easily with frost-seeding, with or without passing animals (sheep) over seeded areas for hoof action to press in the seed. But I haven't been able to establish any grasses (perennial ryegrass, matua prairie grass, tall fescue, Kentucky bluegrass) using the same method. It probably indicates that existing grasses are well adapted to growing conditions and management, and are very competitive against seedlings of new grass species.

Voisin was right! Pasture renovation won't make up for defective grazing management, and usually isn't necessary for improving pastures.

To improve your pastures follow these steps:

1. Don't plow, cultivate, or kill your pasture sod.

2. Test your pasture soils, and correct major soil fertility and pH problems.

3. Subdivide the pasture into as many paddocks as you think necessary to get the recovery periods needed for your area.

4. Follow the Voisin controlled grazing method as closely as possible, but make changes that seem necessary to adapt it to your local conditions.

5. Pastureland that is overgrown with brush and weeds can be improved quickly and inexpensively by applying the technique of mob-stocking. Fence in small areas of the rough land and confine large numbers of nonproducing animals (e.g. dry cows, goats, or ewes) to each small area until they have eaten all of the leaves that they can reach. The areas must be small enough so that they can be grazed completely in 24 hours or less. Every 2 to 3 days return the animals to better pasture for 1 to 2 days so that they don't become stressed for nutrients. Water and salt must be available at all times to the animals.

Begin grazing early in spring just after leaves form on the brush, and graze repeatedly every time the leaves regrow during the season. Before beginning to graze, carefully check the area to make certain that no poisonous plants are present that could harm your livestock when grazed. If there are poisonous plants, remove them from the area. Mob-stocking can change rough land into productive pasture in 2 years or less. This change will be even quicker if you can mow the area with a rotary brush-cutting mower, either in early spring or in midsummer when weedy plants are forming flower buds.

6. Aerate soils compacted by cattle treading.

7. To introduce new plant species or cultivars into pastures out of necessity or curiosity (e.g. perennial ryegrass, prairie grass, birdsfoot trefoil, New Zealand white clovers), try frost-seeding in early spring just before or after the snow melts. Remember to inoculate clover seed. Large-seeded grasses such as prairie grass need to be drilled in after the soil thaws and dries.

8. Be patient. Allow the pasture environment time to recover from the 100 or more years of abuse that it has endured.

9. Lay back and enjoy your healthier, happier, more profitable farm, with its greener pastures on your side of the fence!

REFERENCES

Clark, E.A. 1990. The principles and tools of intensive pasture management systems. Proceedings Field Crops Expo 90, Acadia University, Wolfville, Nova Scotia. July 15-18.

Davies, A., W.A. Adams, and D. Wilman. 1989. Soil compaction in permanent pasture and its amelioration by slitting. Journal of Agricultural Science, Cambridge. 113:189-197.

Frame, J. 1988. Pasture management in Europe. Proceedings Pasture Management Workshop "Improved Production for Your Pasture", Nova Scotia Agricultural College, Truro, Nova Scotia.

Johnstone-Wallace, D.B. 1945. The principles of pasture management. Proceedings of the New York Farmers 1944-1945.

Jones, M. 1933. Grassland management and its influence on the sward. Journal of the Royal Agricultural Society of England. 94:21-41.

Kendall, D. 1988. Nature's no-till pastures. The New Farm. May/June. p. 16-18.

Traupman, M. 1990. Mechanical "earthworms". The New Farm. March/April. p. 12-13.

Voisin, A. 1959. *Grass Productivity*. Island Press, Washington, D.C. 353 p.

Index

296

ORDER FORM

Arriba Publishing
213 Middle Road
Colchester, VT 05446
Telephone 802/878-2347

Please send me Greener Pastures On Your Side Of The Fence: Better Farming With Voisin Grazing Management (2nd Edition) by Bill Murphy @ $19.95.

I understand that I may return the book for a full refund if I'm not satisfied.

Enclosed for   book(s) is _____

(Vermonters please add 5% sales tax: $1.00/book) _____

Shipping (Book Rate): $1.50 per book _____
I can't wait 2-3 weeks for Book Rate mail; here's
$3.00 per book for First Class Mail: _____

Total enclosed _____

Name: _____

Address: _____

_____ ZIP _____

ORDER FORM

Arriba Publishing
213 Middle Road
Colchester, VT 05446
Telephone 802/878-2347

Please send me Greener Pastures On Your Side Of The
Fence: Better Farming With Voisin Grazing Management
(2nd Edition) by Bill Murphy @ $19.95.

I understand that I may return the book for a full refund if
I'm not satisfied.

Enclosed for   book(s) is _____

(Vermonters please add 5% sales tax: $1.00/book) _____

Shipping (Book Rate): $1.50 per book _____
I can't wait 2-3 weeks for Book Rate mail; here's
$3.00 per book for First Class Mail: _____

Total enclosed _____

Name: _____

Address: _____

_____ ZIP _____

ORDER FORM

Arriba Publishing
213 Middle Road
Colchester, VT 05446
Telephone 802/878-2347

Please send me Greener Pastures On Your Side Of The
Fence: Better Farming With Voisin Grazing Management
(2nd Edition) by Bill Murphy @ $19.95.

I understand that I may return the book for a full refund if
I'm not satisfied.

Enclosed for   book(s) is _____

(Vermonters please add 5% sales tax: $1.00/book) _____

Shipping (Book Rate): $1.50 per book _____
I can't wait 2-3 weeks for Book Rate mail; here's
$3.00 per book for First Class Mail: _____

Total enclosed _____

Name: _____

Address: _____

_____ ZIP _____